BRITA

**HODDER AND STOUGHTON**
LONDON SYDNEY AUCKLAND TORONTO

The British Gas Corporation's coastal terminal at
Theddlethorpe, Lincolnshire. (*Photo: British Gas*)

# BRITAIN
# in Colour

AP Fullagar & HE Virgo

*Cover photographs*
*Front: Patrick Bailey*
*Back: British Steel*

Fullagar, Ann Patricia
  Britain in colour.
  1. Great Britain – Description and travel
  I. Title  II. Virgo, Hugh Edward
  914.1      DA632
ISBN 0–340–19179–1

First published 1978

Printed in Great Britain for
Hodder and Stoughton Educational,
a division of Hodder and Stoughton Ltd,
Mill Road, Dunton Green, Sevenoaks, Kent by
Fletcher and Son Ltd, Norwich

# PREFACE

*Britain in Colour* is a systematic study of the geography of Britain, intended for use with the upper forms of secondary schools.

*Britain in Colour* differs from the preceding volume *The British Isles* in its approach to the subject matter, but the essential format of the series remains the same. The text is accompanied by line drawings, which we have tried to keep clean and informative. The use of colour has helped facilitate this aim besides enhancing the quality and value of the accompanying photographs. As usual the questions and exercises are an integral part of the book and are devised to make the students observe, think and investigate.

Our fundamental aim is to provide teachers and students with a text book to assist them in the study of Britain in the limited time that is available for the subject. The book begins by looking at aspects of the physical landscape of Britain and its climate. These factors, together with details of soils and human activity, are considered in relation to agriculture and forestry. Energy plays a vital part in our lives and this subject merits detailed study before leading on to a section on industry. The final parts of the book deal with transport, tourism, population and settlement.

We wish to thank all the people who have helped in the production of this volume, particularly our editor, and to acknowledge our debt to the writers of the numerous books and articles which we have consulted and the various organisations and individuals who have provided material for illustrations.

# CONTENTS

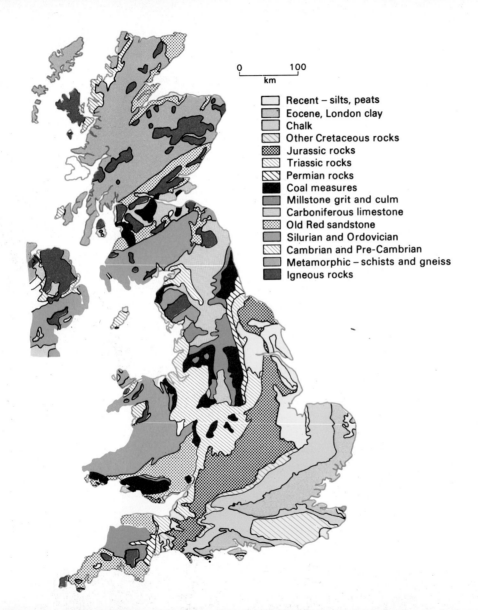

Conditions in the British Isles during the various geological periods.

| | Recent – silts, peats |
| | Eocene, London clay |
| | Chalk |
| | Other Cretaceous rocks |
| | Jurassic rocks |
| | Triassic rocks |
| | Permian rocks |
| | Coal measures |
| | Millstone grit and culm |
| | Carboniferous limestone |
| | Old Red sandstone |
| | Silurian and Ordovician |
| | Cambrian and Pre-Cambrian |
| | Metamorphic – schists and gneiss |
| | Igneous rocks |

0    100
km

**Hadrian's Wall and the Great Whin Sill, Northumberland.** (*Photo: Patrick Bailey*)

## Section 1
## ROCKS AND SCENERY

There were four main types of scenery in Britain when man first settled here.

(1) In upland areas the landscape most nearly resembled that which we see here today: moorland with heather, bracken and coarse grass.

(2) Vast stretches of the lowlands were marsh. These included not only the Fenlands and central Ireland but much of Lancashire, Cheshire, the lower Trent Valley and Somerset as well as more local areas of clay vale in the Midlands and South East England.

(3) Elsewhere in the lower parts of Britain the general scene was one of thick, deciduous forest.

(4) Open grassland country was largely restricted to the permeable uplands. Chalk areas were particularly attractive for settlement.

The natural vegetation was a response partly to the type of rock and partly to the climate. This response has been modified by relief and the effectiveness of the drainage.

Consider the two photographs here. One scene shows a man-made environment with the influence of nature, seemingly, far removed. The other appears to be an attractive scene unspoilt by man and yet the lake is artificial, formed behind the Brianne Dam in central Wales; the woodlands are plantations set out by the Forestry Commission while scenic roads have been constructed for the tourist to enjoy the lovely scene man has created.

Nevertheless, the rocks in a particular region have often stamped a distinctive character on the scenery and had a decisive influence on man's response. In the following section we are primarily concerned with the influence of certain major groups of rocks on the British landscape.

## ROCK TYPES

There are three main groups of rocks classed according to the way they were formed: igneous, sedimentary and metamorphic.

### Igneous rocks

Igneous rocks were once molten material known as magma. This was either intruded into the crustal layers of the earth from below or extruded on to the surface.

Intrusions of magma cooled slowly from a molten condition and the resulting plutonic rocks, as they are called, have large crystals. Granite is an example of this type of igneous rock. Where the intrusions were more

restricted, along joints and bedding planes, to form dykes and sills, an intermediate group of rocks known as hypabyssal were formed; dolerite is an example. Volcanic rocks are formed from magma which poured out on to the earth's surface and cooled rapidly so that crystals are small: basalt is an example.

## Sedimentary rocks

Sedimentary rocks are of two types: those formed from older rocks which have been eroded, deposited in layers and reconsolidated; and those formed from the organic remains of plants and animals. In both cases the deposits may have built up under water or on land. The division between each layer is known as a bedding plane, while the joints at right angles to this plane developed as the sedimentary layer was compacted.

### Classification of sedimentary rocks

(1) Mechanically formed. Derived from the erosion of older rocks and redeposited.

a. Sands, grits and sandstones with quartz the most common mineral in this group. Large rounded fragments cemented together are known as conglomerates; angular fragments when cemented are known as breccia.

b. Fine muds forming clays and mudstones are made of very fine quartz and mica particles.

(2) Organic rocks, the remains of living matter.

(*Photo: Aerofilms*)

*a.* Limestones including chalk formed from the skeletons and shells of sea creatures.

*b.* Peat, lignite and coal formed from plant remains, carbon being the main element.

*c.* Chemical deposits, many formed from the deposition of salts carried in solution. Gypsum for example was deposited in this way.

## Metamorphic rocks

Metamorphic rocks are those which have been changed from igneous or sedimentary rocks by intense heat or pressure. Metamorphism can take place over a large area when earth movements create enormous pressures within rocks. This process, known as regional metamorphism, has changed the shales into slates and schists, the sandstones into quartzites, and granite into gneiss over large stretches of the Highlands of Scotland. Local metamorphism is more likely to be caused by heat from an intrusion of molten magma affecting surrounding rocks. Marble, with its characteristic streaky appearance, is an example of this type.

# FOLDING AND FAULTING

If you study figure 1.1 you will notice that Britain has been affected by a series of earth movements separated by long periods of relatively peaceful conditions. During the mountain building periods, as they are called, great pressures and tensions were set

## The age of rocks

| Number of millions of years ago | Era | Period | | Earth movements | Events and rock types formed |
|---|---|---|---|---|---|
| | Quaternary | Recent | | | Submergence of continental shelf |
| 2 | | Pleistocene | | | Ice age |
| | Tertiary | Pliocene | | Alpine | Shelly sands and gravels |
| | | Miocene | | | |
| | | Oligocene | | | Clays (e.g. London Clay), sands |
| 65 | | Eocene | | | and pebbles in shallow water |
| | Secondary or Mesozoic | Cretaceous | | | Chalk in the upper part. Greensandstones and clays in lower. Seas deepened during the period. |
| | | Jurassic | | | Seas covered much of England and Wales. Alternate layers of limestones, sandstones and clays. |
| 280 | | Triassic | Keuper | | Desert conditions. Keuper marls. |
| | | | Bunter | | Bunter Sandstones |
| | Primary or Palaeozoic | Permian | | Hercynian or Armorican | Magnesian limestone |
| | | Carboniferous | | | Coal measures- shallow lagoons and swamps |
| | | | | | Millstone grit- deltas- coarse sediments |
| 345 | | | | | Carboniferous limestone- deep sea |
| 395 | | Devonian | | Caledonian | Marine sediments. Old Red Sandstone deposited in fresh water lakes |
| 440 | | Silurian | | | Mostly marine sediments with intrusions and extrusions. |
| 500 | | Ordovician | | | Resistant grits and sandstones |
| 570 | | Cambrian | | | with metamorphics such as slate |
| | Eozoic | Pre-Cambrian or Archaen | | Charnian and others | Very ancient rocks including gneiss, schist, lava and quartzite |

up, causing folding or bending of the outer layers of the earth's crust. These movements were accompanied by faulting and massive intrusions of magma into the upper layers of the earth's crust or extrusions over the surface of the earth.

The old fold mountains of Caledonian times have been subjected to such a long period of erosion that they now appear as small remnants or stumps of what must have been great mountain ranges. The Alpine earth movements brought only ripples to Great Britain but nevertheless these ripples started an erosion cycle which has given the south east a distinctive scenery of scarps, dips and vales.

Pressure and tension can cause fractures in the earth's crust. Further pressure can cause movement along the line of the fracture. Faulting of this nature occurs more frequently in resistant older rocks which do not 'give' to pressure by folding.

Describe this feature which occurs on the Pembrokeshire Coast near Little Haven.

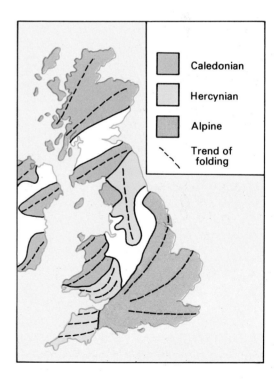

**Figure 1.1  Simplified structure of Great Britain.**

**Figure 1.4  Major fault systems in Scotland.**

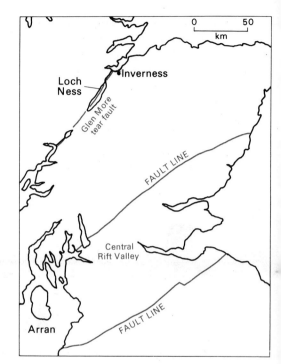

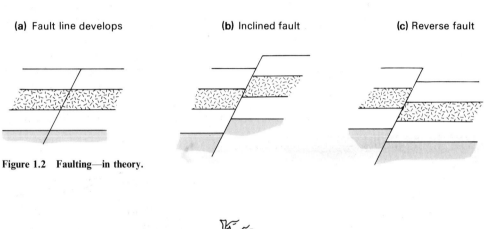

**(a)** Fault line develops  **(b)** Inclined fault  **(c)** Reverse fault

**Figure 1.2  Faulting—in theory.**

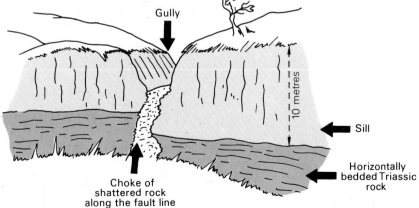

Gully

10 metres

Sill

Horizontally bedded Triassic rock

Choke of shattered rock along the fault line

**Figure 1.3  Faulting—in practice.**

# IGNEOUS ACTIVITY AND ASSOCIATED SCENERY

The section and map (figures 1.5 and 1.6) show a portion of Arran, an island in the estuary of the Clyde. If you were to walk along the line of the section you would cross a number of different outcrops of rock. Near the east coast is Permian sandstone, then Old Red sandstone followed by ancient Dalradian schists. West of the schists and making up most of the northern half of the island is granite.

In Arran this roughly circular outcrop of granite, which today covers 155 square kilometres of the surface, must have forced up the overlying rock into a giant blister. Then the molten material cooled and solidified slowly enough to form well defined crystals of quartz, felspar and mica, the crystals of granite. Over a long period the agents of denudation (mainly ice and water) exposed the underlying granite.

Arran was subjected to later igneous intrusion. This time not a massif injection of magma but many small injections. In the granite areas the dykes have tended to be less resistant than the surrounding rock; consequently, they have been eroded more rapidly to form deep gullies. The dykes and sills intruded into sedimentary rock have generally proved more resistant than the surrounding rock: these features are particularly obvious around the coast. Sometimes an intrusion has baked the surrounding sedimentary rock to make it locally more resistant.

These features of igneous activity can be seen over many parts of Highland Britain. Some igneous rocks, which were extruded over the surface of the land as a molten lava and cooled rapidly, took on a columnar structure as in the Giant's Causeway on the north coast of Northern Ireland. This particular extrusion of basalt makes up the Plateau of Antrim. Other basaltic plateaux occur on Mull, Skye and other islands off the

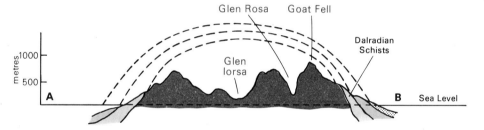

Figure 1.5   Section across northern Arran.

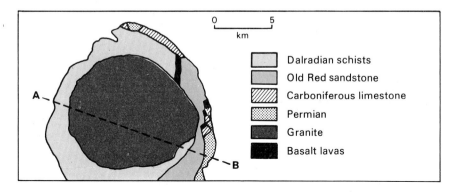

Figure 1.6   Simplified solid geology of northern Arran.

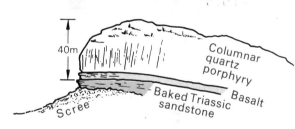

Figure 1.7   Sketch of Drumadoon Head—a sill.

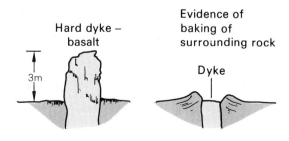

Figure 1.8   Sketches of dykes near Drumadoon Head.

7

west coast of Scotland, the columnar structure being seen again at Fingal's Cave on Staffa. There have been volcanoes but evidence for them is difficult to trace because of the effects of long periods of erosion. Remnants of the plugs of volcanoes are apparent; for example, as the hills on which Edinburgh and Stirling castles stand.

The more subdued, unglaciated granite moorlands of Devon and Cornwall have a distinctive scenery. Here granite bosses have been denuded to form rolling country with the crests often dominated by outcrops which consist of rounded blocks of granite rising from a weathered mantle of granite debris. In some areas the decomposition of the felspars in the granite has resulted in large accumulations of kaolin (or china clay) round the edges of granite bosses.

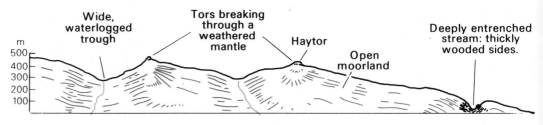

**Figure 1.9 Dartmoor—sketch of a portion of the eastern skyline.**

Identify the features illustrated and explain how they have been formed. They are a soft and a hard dyke, a massive sill (all taken along the west coast of the Isle of Arran) and a tor on Dartmoor but they do not occur in this order. (*Photos: Howard Thomas*)

8

# PENNINE SCENERY

A look at the geological map (figure 1.11) shows that the Pennines are made up of two main groups of rock, carboniferous limestone and millstone grit.

## The Derbyshire Peak District

Between Ashbourne and Buxton is a wide, undulating plateau. The land surface is dry and there are very few watercourses. The few trees to be found are concentrated in small clumps in hollows, which are also often the sites of the scattered farm buildings. Otherwise the open plateau is criss-crossed by a patchwork of dry stone walling. Small circular ponds, called dew ponds, dot the countryside; these are mostly man-made, being floored with concrete and used for livestock. Numerous limestone quarries occur: many have been abandoned but there are large workings particularly near Buxton, providing raw materials for metallurgical and chemical industries as well as for the manufacture of cement.

On the edge of the plateau there is far more rugged scenery with streams occupying steep-sided valleys, the sides being thickly wooded. Along these valley sides great blocks of the light grey Carboniferous limestone outcrop.

## The limestone areas of the middle Pennines

In this region the country is more rugged with tracts of wild moorland and great scars where the limestone outcrops.

Carboniferous limestone is a non-porous rock which, whilst it resists mechanical erosion by weathering, is dissolved by rainwater containing carbon dioxide. It is a well-jointed permeable rock and a large proportion of rainwater penetrates its joints and runs underground leaving a dry surface.

Over most of the Pennines the limestone strata are repeatedly interspersed with layers of shale and other rock. However, in this central Pennine region in the Ingleborough–Malham area, a thick layer of Great Scar limestone has developed a karst scenery including limestone pavement, pot holes and swallets, while underground there is an intricate system of caverns and channels where the running water has eaten into the joints and bedding planes.

## Millstone grit between the industrial areas of Yorkshire and Lancashire

If you cross from the limestone plateau of Derbyshire to the millstone grit area there is a significant contrast in scenery, most obvious being the change in colour from the light grey walling and green fields on the limestone to the sombre, dark colours of the grit.

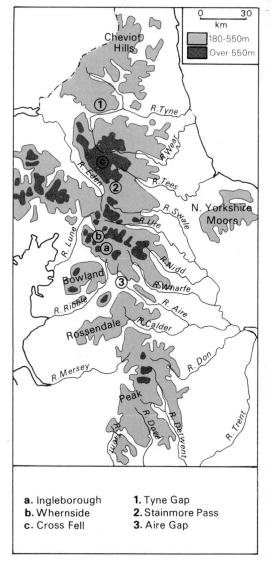

| | |
|---|---|
| **a.** Ingleborough | **1.** Tyne Gap |
| **b.** Whernside | **2.** Stainmore Pass |
| **c.** Cross Fell | **3.** Aire Gap |

**Figure 1.10   The Pennines—physical.**

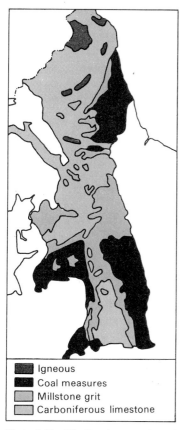

**Figure 1.11 The Pennines—geology simplified.**

- ■ Igneous
- ■ Coal measures
- ▨ Millstone grit
- ▨ Carboniferous limestone

**Carboniferous limestone plateau above Malham Cove, Yorkshire.** (*Photo: Geological Survey—Crown copyright*)

Millstone grit is an impermeable rock, and this accounts for the abundant surface drainage pattern. Drainage is however often poor and there are tracts of peat bog which have a vegetation cover of cotton grass. Elsewhere the dominant vegetation cover is heather. These damp, acid soils, supporting such a coarse vegetation, do not favour farming, and the heavy rains and industrial smoke in the central region have so soured the land that improvement would be difficult. However, the upper portions of many of the narrow valleys have been dammed and the resulting reservoirs provide water for the densely populated neighbouring industrial regions.

a

b

(a) Upper Glen Rosa; (b) Cluster of moraines; (c) Looking up the valley from the mouth of Glen Rosa.

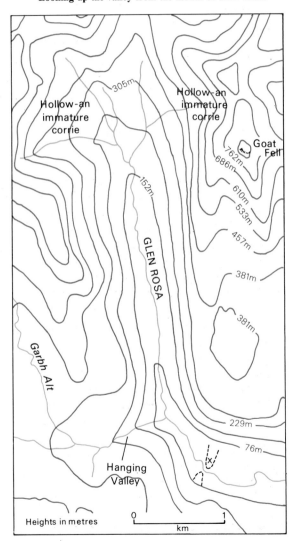

Hollow-an immature corrie

Hollow-an immature corrie

Goat Fell

305m

762m
686m
610m
533m
457m
381m
381m
229m
76m

152m

GLEN ROSA

Garbh Alt

Hanging Valley

X

Heights in metres

0      1
km

**Figure 1.12** Glen Rosa.

## GLACIATION IN UPLAND AREAS

The photograph and plan (figure 1.12) are of a Scottish glen, Glen Rosa. This spectacular trough is occupied by a small stream. Note the smooth sides of the glen above which there is a break in slope leading on to the rugged and broken uplands. There are other clues which can be found which help to explain this misfit.

At 'x' on figure 1.12, a low mound between four and five metres high extends across the valley. The present stream has cut into this to reveal a river cliff face 3·5 metres high composed of loosely consolidated stones and coarse grit. There are remnants of other such mounds lower down the glen.

These features are all characteristic of a glaciated valley. The mounds are moraines which mark the limits of a glacier at various stages in its retreat; they are known as recessional moraines.

High up on the sides of the upper portion of the valley there are hollows. Here snow collected and was compressed into ice and the resultant glacial erosion deepened the depressions into shallow arm-chair hollows or corries.

These and other glacial features are characteristic of other mountainous areas in Scotland, North Wales and the Lake District of Cumbria. During the Ice Age, when the ice

◀ Section of a moraine exposed by river erosion.

13

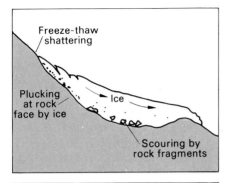

Freeze-thaw shattering

Plucking at rock face by ice

Ice

Scouring by rock fragments

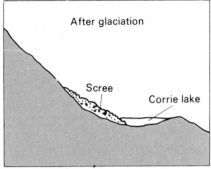

After glaciation

Scree

Corrie lake

**Figure 1.13  The formation of a corrie.**

was at its maximum, these uplands were covered with thick layers of ice taking the form of ice caps. At this time the ice must have acted more as a protection from erosion. At the end of the Ice Age, when there was an incomplete cover, the ice moved in streams away from the collecting areas. Although we have no glaciers in Britain to-day, from the study of glaciers in other parts of the world we can piece together how the landscape of our own glaciated areas evolved.

The photograph shows the collecting ground, an extensive upland hollow which has a permanent snow cover. As this snow accumulates it is compressed into glacial ice: in this case the ice builds up to such a mass on a sufficiently steep slope that a large enough downward force is created to begin a movement of ice downstream. One of the most potent effects of glacial erosion occurs when the glacier plucks rocks from the bed of the valley and these rocks adhere to it. It is these rocks, embedded in the ice, which are a powerful abrasive force grating away at the sides and floor of the valley along which the glacier passes.

We have referred to the shallow hollows on the side of Glen Rosa. More spectacular hollows occur, for example, in the Lake District. Towards the end of the Ice Age the deeply entrenched valley glaciers were becoming spent. Higher up, and particularly on the north-east sides of the uplands on which the sun never shone directly, the hollows were liable to collect much snow which in turn was compressed into ice.

After the ice has retreated the floor of a glaciated valley is normally very uneven. It is therefore not surprising that the hollows were frequently occupied by lakes. These lakes were far more extensive than they are today.

After the Ice Age, fast-flowing streams, many of which tumble down the steep-sided valley from tributary valleys now left 'hanging' above the main valley, brought large

PLAN

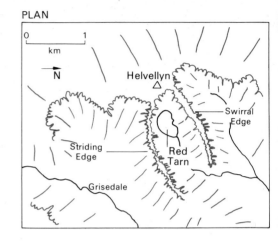

0    1
km

N

Helvellyn

Swirral Edge

Striding Edge

Red Tarn

Grisedale

FIELD SKETCH

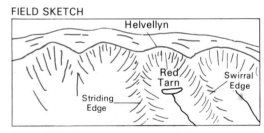

Helvellyn

Red Tarn

Swirral Edge

Striding Edge

**Figure 1.14  Corries and aretes—Helvellyn.**

quantities of sediment which has been deposited as alluvial flats, fans and deltas.

# GLACIATION IN LOWLAND AREAS

The mountainous areas of Britain have many easily identifiable glacial features.

**Arete and corrie on the flanks of Snowdon.** (*Photo: Aerofilms*)

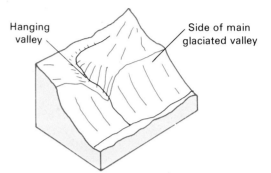

**Figure 1.15    A hanging valley.**

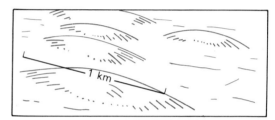

**Figure 1.16    Sketch of drumlin country.**

**Figure 1.17    The extent of glacial boulder clay over the Midlands and eastern England.**

The glacial features of lowland areas are far more extensive but less obvious. For example, study the map (figure 1.17), which shows the extent of boulder clay deposits in the English Midlands. Boulder clay is a mixture of rock fragments and clay which was dumped unevenly across the countryside when the ice sheet decayed.

There are other more obvious features. Drumlins consist of glacial debris, gravels, rock fragments and clay which have been shaped under the ice-sheet into a series of rounded elongated hills. Lowland moraines, such as the York and the Escrick, cross the Vale of York and mark stages in the retreat of a thin ice-sheet at the end of the Quaternary glaciation.

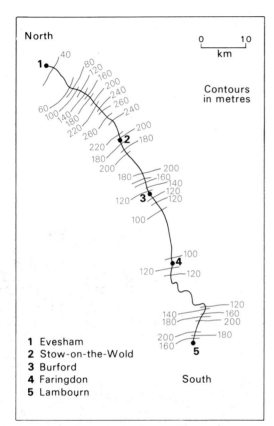

**North**

0 ___ 10
km

Contours
in metres

1 Evesham
2 Stow-on-the-Wold
3 Burford
4 Faringdon
5 Lambourn

South

**Figure 1.18   The route from Lambourn to Evesham.**

# SCARPS AND VALES

Study figure 1.18, showing a portion of
southern England. In particular pay attention to the spacing of the contours along the
road from Lambourn to Evesham. Notice
how the contours north of Lambourn at first

indicate rising land as you travel northwards. Then there is an abrupt descent where
the contours are close together. The slope is
succeeded by a flat area marked by an
absence of contours. The flat area, or vale, is
broken by a low ridge on which Faringdon is
sited. From Lechlade the sequence of slopes
is repeated: a gentle but more broken slope
followed by a steep one and then a flatter
area around Evesham.

Figure 1.19 is the result of converting this
map into a block section and marking in the
rock types crossed. The section shows the
relationship between the beds of sedimentary
rock dipping very gently to the south and
south-east and the relief features. The more
resistant rocks form the hill country and
their gentler slopes follow the dip of the rock
strata, the steep slope marking the outcrop
of the resistant strata.

This scarp and dip feature is characteristic
of the relief of south and east England. The
Alpine fold mountain-building period, in
which most of the great mountain ranges of
the world have their origin, brought only
gentle folding to the rocks of southern and
eastern England which lay at some distance
from the zone of intense folding. As a result
of this slight Alpine folding there are two
almost parallel belts of hill country, one of
Jurassic limestone and the other of chalk,
running from Yorkshire to the Dorset coast
with other chalk hills radiating from
Salisbury Plain to the east and south-east.

To return to our example, the Berkshire

Downs, which are of chalk, form the southern edge of our section and rise to between
200 m and 270 m above sea level near the
north-facing scarp.

## The chalk

Chalk is a soft white limestone—a fairly pure
form of calcium carbonate. Although it has
few deep joints or cracks, it has minute pores
and thread-like fractures so that rainwater
passes through, very slowly. Chalk is sufficiently permeable to be devoid of surface
drainage. The overall relief outline is one of
smooth, rounded curves with convex slopes
dominating. As the photograph shows, the
soil that develops on the chalk is thin; it
leads to the development of a short tough
grass with scattered patches of box, yew and
juniper. There are clumps of beech, whose
roots are shallow but radiate out for considerable distances.

The scarp face of the Berkshire Downs is
broken by short, steep-sided, dry valleys with
an abrupt upper end. These are known as
combes. On the dip slope the dry valleys are
longer and shallower with tributary valleys
leading off. At the foot of both the scarp and
dip can be traced a spring line; this is near
the junction of the chalk with the underlying
impermeable clay (see figure 1.20).

The level of the water table varies depending on the season of the year. After a rainy
spell it will be slightly higher. When it is very
high temporary streams, known as bournes

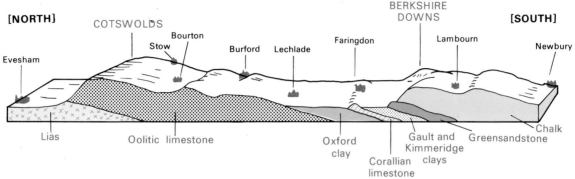

[NORTH]
COTSWOLDS
Bourton
Stow
Burford
Lechlade
Faringdon
BERKSHIRE DOWNS
Lambourn
Newbury
[SOUTH]
Evesham

Lias
Oolitic limestone
Oxford clay
Corallian limestone
Gault and Kimmeridge clays
Greensandstone
Chalk

**Figure 1.19   Section from Newbury to Evesham.**

or winterbornes, may flow in the 'dry' valleys for several weeks. The valleys in the chalk must have been formed by surface streams in the past when the water level was higher.

On top of the chalk there are patches of clay-with-flints. This capping is an accumulation of an insoluble residue left after chalk has dissolved and been removed. On clay-with-flints the soil is damper, heavier and deeper so that more varied vegetation including oak and hazel develops.

The photograph shows a scene on the dip slope of the chalk. Notice that much of the land is under the plough, which is not true of the scarp slope: why is there this difference? The fields are large as are the farms themselves, many being over 400 hectares. The light soils are particularly suited to barley though wheat, oats and root crops are also grown. The land is normally rested from crop production after a few years. There is

grass—under a ley perhaps for three or four years during which cattle and sheep graze the pasture.

## Other chalk areas

This study of Berkshire chalk brings out features which occur in other chalk areas of the country; but there are differences. For example on the North Downs and the Chilterns much of the dip slope has been invaded by suburbia; in some areas quarrying is significant. Chalk is not a useful building stone so why do they quarry it? The chalk scarp of the Chilterns is very extensively clothed by woodland, especially by beech.

The gaps in the chalk have proved of great significance for the location of settlement

**Chalk.** ▶

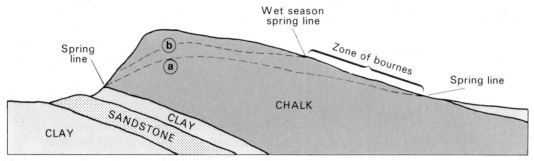

**Figure 1.20 The water table in chalk.** (a) The normal level of the water table. (b) The level of the water table after a period of heavy rainfall.

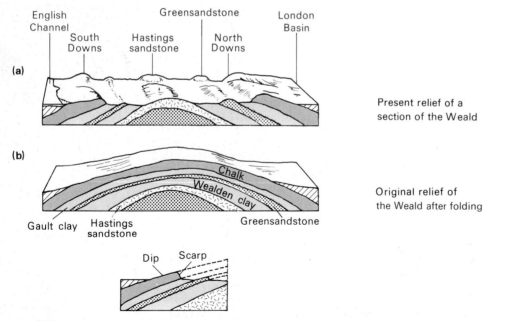

**Figure 1.21 The Weald of south east England.** (a) The general relief today. (b) The anticline before denudation.

and the communication pattern. Roads and railways have tended to use the gaps rather than negotiate the scarp slopes and this has encouraged the growth of settlement at the mouth of the gap.

## Jurassic limestone

The other more resistant rock in this section is Jurassic limestone, which forms the Cotswolds. This rock is, like chalk, calcium carbonate, but it is older, harder and more jointed. Consequently the scarp slope is more rugged than that of the chalk and there are outcrops of bare rock.

The overall scene on the Cotswolds is one of less rounded, more angular relief forms than on the chalk. Although the rock is relatively permeable and the upper surface of the limestone is dry there is some surface drainage here, the watercourses occupying deeply etched valleys cutting into the dip slope. Jurassic limestone is a valuable building stone and much of the difference in the scenery of chalk and Jurassic limestone is due to use of the limestone to form dry-stone

**South Downs. 1.** Draw a sketch section across the area from left to right passing through the village of Fulking. Label the dip slope, scarp slope, village and clay vale. **2.** (a) Explain how the Chalk scarp has developed and why springs occur at its foot; (b) Compare the land utilisation on the Chalk downs with that on the clay vale (to the right of the road); (c) Give as many reasons as possible why the village of Fulking has grown up here. (*Photo: Aerofilms*)

Jurassic limestone.

walling and for buildings, which have a distinctive, warm, yellow-to-brown colour.

Over this stretch of the Cotswolds the land use is similar to that on the Berkshire Downs. Both were sheep farming areas but now much of the land is under the plough, although there remains a greater proportion of permanent pasture where the relief is more broken.

## The claylands

Between the chalk and the Jurassic limestone is a broad, undulating plain broken in two by a slight rise where Corallian limestone outcrops. The Oxford, Kimmeridge and Gault clays which floor this plain are finely grained, soft rocks which are easily eroded and very retentive of moisture. Clay does not have joints or bedding planes and once it is saturated it is impervious. Consequently there is much surface drainage. The rivers have built up alluvial coatings and here the relief is very flat indeed.

These claylands were once covered by thick, deciduous woodland interspersed with waterlogged marshy areas on the alluvial tracts. Over the centuries these lands have been cleared and now only the hedgerows which separate the fields together with numerous small copses give the appearance of a well-wooded landscape.

On claylands there have developed much heavier, thicker soils far more retentive of moisture. Such soils will produce a far richer

grassland than that on chalk or limestone so that here there is much more permanent grassland. Individual farmers specialise, some on beef cattle, some on dairying and, less often, on crops.

There is a far more even distribution of rural settlement over these claylands. However, the areas which are liable to be water-logged are avoided and slight elevations provide the sites for many of the villages.

**The Cotswolds scarp.** In what way does this scene differ ▲ from that of the South Downs? Explain why these differences occur. (*Photo: Geological Survey—Crown copyright*)

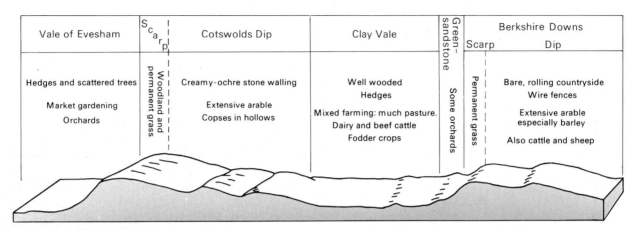

| Vale of Evesham | Scarp | Cotswolds Dip | Clay Vale | Green-sandstone | Berkshire Downs |
| | | | | | Scarp Dip |
| Hedges and scattered trees | Woodland and permanent grass | Creamy-ochre stone walling | Well wooded | Some orchards | Bare, rolling countryside |
| Market gardening | | Extensive arable | Hedges | | Wire fences |
| Orchards | | Copses in hollows | Mixed farming: much pasture. | Permanent grass | Extensive arable especially barley |
| | | | Dairy and beef cattle | | |
| | | | Fodder crops | | Also cattle and sheep |

Figure 1.22   Land use along the line of the transect.

## EXERCISE

Write an account which describes and explains the distinctive characteristics of chalk, Jurassic limestone and clay scenery. Include in your account a consideration of field boundaries, land use, uses made of each rock, and drainage patterns.

## The river system of the Upper Thames Basin

Study figure 1.23, which illustrates the development of the drainage pattern in the Upper Thames Basin. Following the Alpine tilting of the strata, all the streams must have flowed with the general dip of the land towards the south or south-east. Streams flowing with the dip of the strata are called

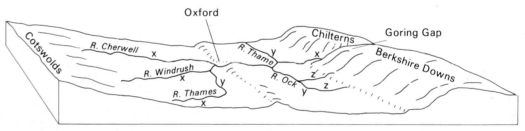

Figure 1.23 Major drainage pattern of the Upper Thames Basin. X—consequent stream; Y—subsequent stream; Z—obsequent stream.

consequent streams. With the wearing down of the land surface the clay vales were formed and were occupied by subsequent streams, flowing roughly at right angles to the consequents. Other smaller obsequent streams flow into these subsequents, against the dip of the strata.

It will be seen that the Thames flows first as a consequent stream on the dip slope of the Cotswolds, then along the Oxford Clay Vale as a subsequent stream before turning south to become a consequent again. Finally it leaves the Upper Thames Basin passing through the Goring Gap between the Berkshire Downs and Chilterns and entering the London Basin.

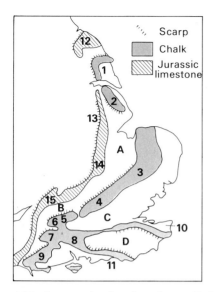

**Figure 1.24 Scarps and vales of south east England.** Identify (a) the uplands 1–15, and (b) the lowlands **A–D**.

With such a trellised drainage pattern, river capture frequently occurs and there are examples of this here. Probably the Cherwell was the original headwater of the Thames and the present Thames above Oxford captured the headwaters of a number of south-flowing streams including the Windrush.

**Wenlock Edge. 1.** Draw a section from left to right across the area shown on the photograph. Mark in scarp and dip slopes. **2.** Where does most of the woodland occur and why do you think this is so? **3.** What types of farming occur in the area covered by the photograph? (*Photo: Aerofilms*)

# RIVER DEVELOPMENT

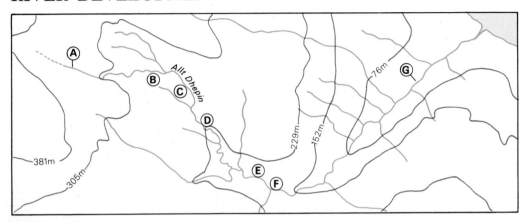

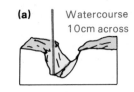

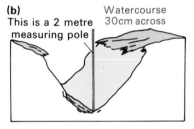

**Figure 1.25  The river study.**

## The mountain stream

Consider the map and diagram (figures 1.25 and 1.26) which illustrate the course of a small river flowing across an upland area in North West Scotland.

In the Highlands of Scotland most rocks are ancient, resistant and impermeable: schists, gneisses and granite are typical. Over much of this area watercourses begin as rivulets trickling from water-saturated hollows. Contrast this with the beginning of watercourses on the edge of chalk areas. At first a rivulet's erosive power is negligible as it winds among the bracken and heather, often hidden from view. As its volume increases there comes a stage when the fast flowing stream is capable of downcutting its bed.

**Figure 1.26  Sketch sections taken at various points along the course of the river.**

(a), (b), (c), (e) and (g) show the development of the valley from near the source of the river to its mouth. Describe and compare the chief characteristics of the sections across the valley at these points. (d) and (f) show two types of waterfall. Why do they occur and how do they differ?

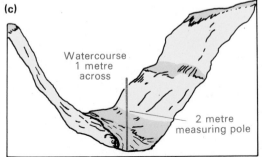

24

**(d)**

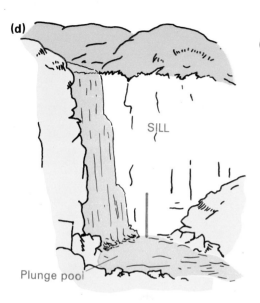

SILL

Plunge pool

**(e)**

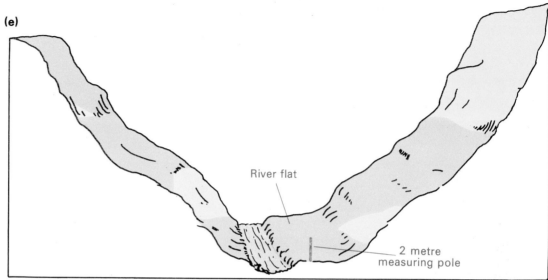

River flat

2 metre
measuring pole

**(f)**

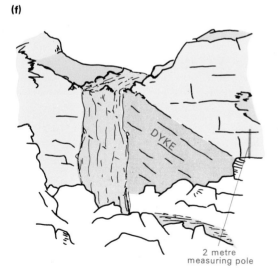

DYKE

2 metre
measuring pole

**(g)**

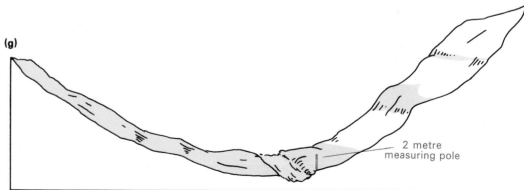

2 metre
measuring pole

25

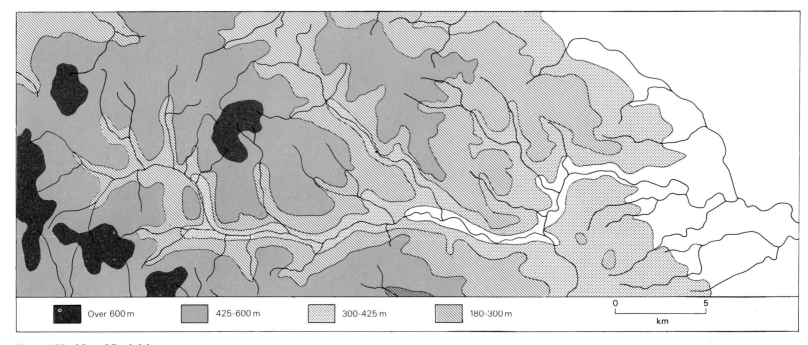

Over 600 m ☐  425-600 m ☐  300-425 m ☐  180-300 m ☐

0   5
km

**Figure 1.27   Map of Swaledale.**

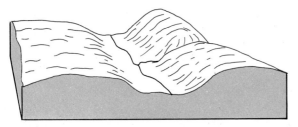

**Figure 1.28   Block section of a portion of Swaledale.**

## The river valley

### EXERCISE

Figure 1.27 is a map of a portion of the Yorkshire Dales. The major valley on this extract is Swaledale. Figure 1.28 is a block diagram of a portion of Swaledale about 20 kilometres from the River Swale's source. From the evidence of the map and the cross section, describe the character of the Swale valley. Pay particular attention to the shape of the valley, the nature of the valley floor and the course of the river. For example, is the river winding between interlocking spurs; is it cutting down into the bedrock; or has it built up an alluvial cover over the bedrock?

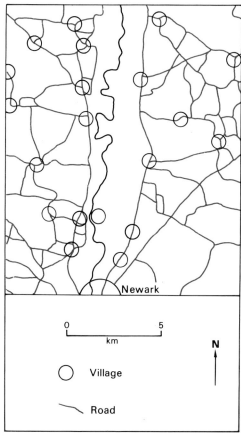

Figure 1.29 Map of a portion of the Trent below Newark.

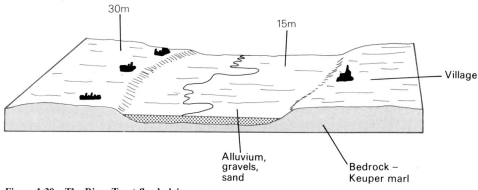

Figure 1.30 The River Trent flood plain.

## The flood plain

The Vale of Trent (see figures 1.29 and 1.30) is underlain by Keuper Marl but this has been masked by thick layers of drift material: clays, sands and silts of both glacial and alluvial origin. The area was a glacial lake at the end of the Ice Age when the ice was retreating, so lake silts make up some of the drift.

Over this vale the River Trent has built up a flood plain as a result of periodic flooding. In the past, before the river banks had been artificially built up, flooding was frequent and extensive.

## COASTAL SCENERY

The photograph, and figure 1.31, illustrate a portion of the west coast of Scotland. The wave-cut platforms at different levels are evidence of fluctuations in sea level. The lowest raised beach at 8 metres above sea level is backed by a cliff face and caves worn by wave action: such features, together with isolated stacks set in grassland, provide further evidence of a fall in sea level. Higher up the traces of raised beaches become blurred because of subsequent weathering and can only be recognised occasionally.

These are features of coasts of emergence and they are found along many stretches of coast around Britain.

Movements have also periodically taken place in reverse, with a rise in sea level in relation to the land. This rise in sea level was on a wide and significant scale after the Ice Age, and the coasts of Devon and Cornwall, South West Wales, the west coast of Scotland and South West Ireland show many examples of drowning of previously low-lying land.

Abandoned sea cliff with raised beach; a scene which is found along many stretches of the west coast of Scotland and Wales.

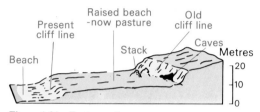

Figure 1.31  Features of emergence.

Figure 1.32 illustrates the drowning of the lower portion of a river valley. The resultant feature is called a ria.

Even more impressive is the drowning of a glaciated valley (see figure 1.33). The resultant feature in this case is called a fjord, and examples of this occur on the west coast of Scotland.

Emergence and submergence of coasts are the results of earth movements or changes in sea level. Such changes can give features of both emergence and submergence along the same stretch of coast. What example mentioned in this section has both types of

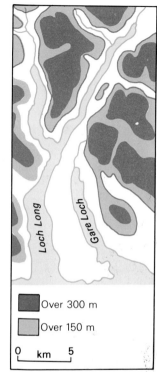

Figure 1.33  A Scottish fjord.

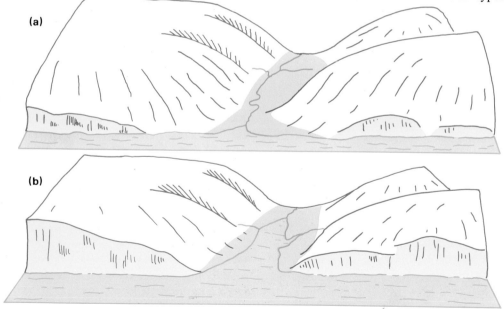

Figure 1.32  The formation of a ria. (a) Before a rise in sea level. (b) After a rise in sea level.

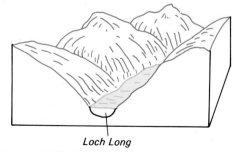

Loch Long

Figure 1.34  Block section of a Scottish fjord.

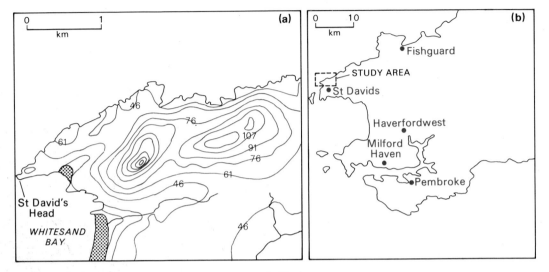

**Figure 1.35    (a) Relief and (b) location maps of St. David's Head.**

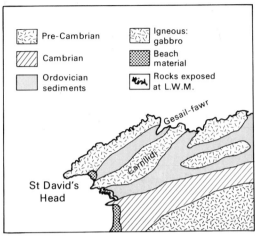

**Figure 1.36    Geology of the St. David's Head area.** What effect does geology have on relief and the character of the coastline?

features? We have already seen that crustal movements which alter the level of the land have frequently taken place. Changes in sea level can be associated with the amount of water covering the earth. At the end of the Ice Age the melting of vast stretches of ice-sheets resulted in a significant rise in sea level.

## Coastal erosion

Figures 1.35 and 1.36 illustrate a coastline which develops where alternate bands of relatively more resistant and less resistant rocks run at an angle to the coast. Initially the less resistant rocks are worn back by marine erosion more rapidly but subsequently the bay which has formed here is protected by the promontories against which the full force of storm waves is expended. A characteristic of the development of coasts is the uneven rate of effective erosion. For much of the time erosion can be negligible but during times of storm, particularly at high tide, massive blocks of rocks can be removed.

Figure 1.35 is a map of a portion of South West Wales that is made up largely of

shales and sandstones with massive injections of magma, which has become the more resistant rock. The coast has been drowned and the resulting inlets, notably Milford Haven, are rias. Subsequently the coast, exposed to the frequent south-westerly storms, has become a most spectacular one of rugged promontories, stacks (which are the stumps of former promontories), cliffs and sheltered bays. It is not surprising that this coast of outstanding natural beauty has been designated a national park.

**St. David's Head.** Study this scene and the geological and physical maps of the area. Then draw a sketch map of the area covered by the photograph marking in the geological boundaries as far as you can. In what ways are the geological changes reflected in the physical geography of the area? (*Photo: Aerofilms*)

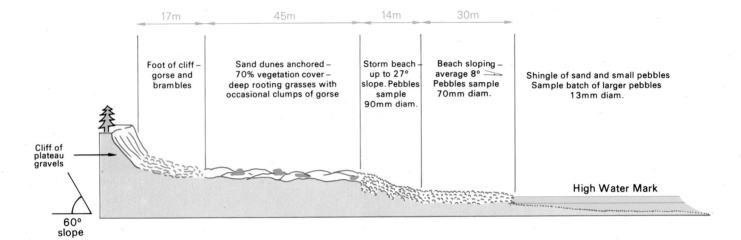

Figure 1.37 labels: 17m, 45m, 14m, 30m

Foot of cliff – gorse and brambles

Sand dunes anchored – 70% vegetation cover – deep rooting grasses with occasional clumps of gorse

Storm beach – up to 27° slope. Pebbles sample 90mm diam.

Beach sloping – average 8° Pebbles sample 70mm diam.

Shingle of sand and small pebbles Sample batch of larger pebbles 13mm diam.

Cliff of plateau gravels

60° slope

High Water Mark

**Figure 1.37 Field section of the beach at Lepe.**
The pebble sample was taken using a one metre square frame and measuring the maximum diameter of each of the surface pebbles. An average was made from these results. For the shingle, however, only a sample collection of twenty-five of the larger stones was taken and an average made.

## Coastal deposition

Study the annotated diagram (figure 1.37) of a beach near Lepe on the Hampshire coast. In such a study the following points are significant.
(1) The degree of slope at different points down the beach.
(2) The size of beach material. This normally varies at different points so that a series of samples must be taken.

(3) Evidence of movement of material along the beach. The difference in height of the beach either side of a groyne would indicate such movement.
(4) The character of the coastline behind the beach. For example, is there a cliff face and if so what form does it take? Is there any evidence of recent wearing back?
(5) Does the material on the beach have any connection with the cliff?

### EXERCISE

Using the evidence of the section and the photograph write an account of the beach under the five sections listed above.

Some beaches have so much debris piled up above the high water mark, that this material masks the coastline and consequently protects it from further marine

erosion. Such a beach is called a storm beach and this is a common feature around the shores of Britain.

Another form of beach where active erosion of the coast is very limited, is where its slope is so gentle that strong breakers, even in times of storm, rarely reach the shoreline.

Although it is possible to recognise beaches where the chief characteristic is one of accumulation and stagnation, most beaches are made up of material in transit. This movement of material is most marked along the lower part of the beach, which is being continually worked by the sea, and is least obvious along the upper beach, where the material is normally larger in size and likely to be moved only by storm waves at high tide.

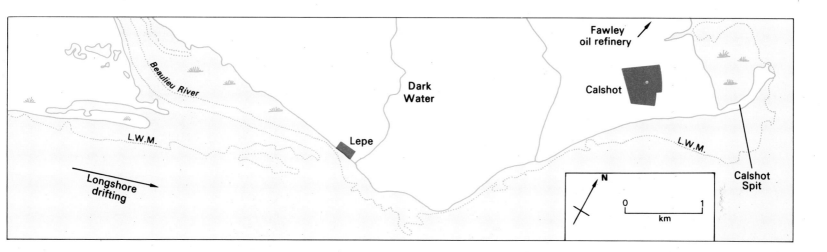

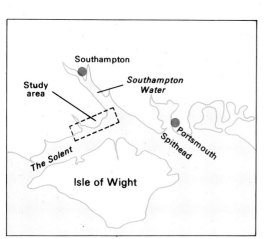

**Figure 1.38   The coast study area in Hampshire.**

**Lepe Beach.** This scene occurs near to the section study. Describe the character of the cliff. What is the purpose of groynes? ▶

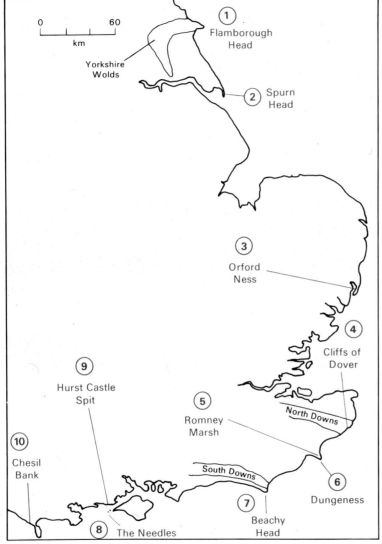

**Figure 1.39  Coasts of emergence and deposition.**
What features on the map are examples of erosion and what are examples of deposition? Explain your decision.

# INFERTILE LOWLAND SCENES

Figure 1.40 shows the extent of six infertile tracts in the lowlands of England. Locate the six descriptions below with the corresponding areas on the map.

(*a*) Plateau gravels and sands of the New Forest, Hampshire. Oak woods occur particularly where there is a clay covering over the gravels. On the sandier areas beech and birch are dominant. There are extensive stretches of heath with heather, gorse and coarse grass, together with coniferous afforestation schemes.

(*b*) The Breckland which stretches across the borders of Norfolk and Suffolk. This is an area covered by a thick layer of outwash sands deposited by watercourses running from the decaying ice-sheet at the end of the Ice Age. This has proved a heartbreak area for farmers because of the hungry soils. Now most of the Breckland has been developed as coniferous plantations by the Forestry Commission.

(*c*) The sandstone outcrops of the Weald of South East England. The most extensive of these are the Forest Ridge, which is made up of Hastings sandstone, and the Upper Greensandstone Ridge, particularly in the Leith Hill area. Portions of this area have been cleared for mixed farming but much remains as deciduous woodland.

(*d*) Sherwood Forest on the Bunter sandstone outcrops in Nottinghamshire. This

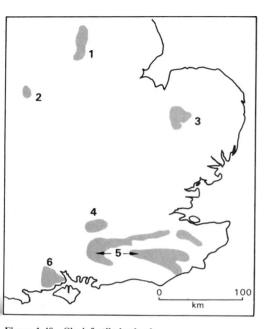

**Figure 1.40  Six infertile lowland scenes.**

**The Forest Ridge of the Weald of South East England.** Describe the scene. Why is it typical of a sandstone outcrop in lowland Britain? (*Photo: Geological Survey—Crown copyright*)

remains a region of infertile land, although inroads into the once continuous stretch of deciduous woodland and heath have been made by farmers. The Forestry Commission has also established coniferous plantations on some of the heathland.

(*e*) Cannock Chase, the plateau area in Staffordshire. The dissected low plateau is made up of Bunter pebble beds and is a sparsely populated area covered by natural and plantation woodlands as well as heath.

(*f*) Bagshot beds in north-west Surrey and east Berkshire. Much of this area of gravels,

which is particularly infertile and covered by birch and heath vegetation, is used by the army for camps, ranges and training grounds and the Forestry Commission has also developed coniferous plantations.

The six regions have a similar geology—sands, gravels and sandstones. The soils are acidic and deficient in soil nutrients. Such lands are marginal for agriculture. By this we mean that with effort and expenditure on heavy applications of organic fertilisers they can be made productive, but there must be the incentive of high returns to make farm-

ing worthwhile. The farms which did exist on the Breckland, for example, have largely been abandoned and occupied by the Forestry Commission. However, these six regions are of significance as attractive open spaces which we can drive to, walk over and picnic on.

The scenes show four distinct aspects of physical geography. Identify and account for each of these aspects. (*Photos: Patrick Bailey*)

# Section 2
# CLIMATE AND WEATHER

(*Photo: Aerofilms*)

*Precipitation* (mm)

|  | J | F | M | A | M | J | Jy | A | S | O | N | D |
|---|---|---|---|---|---|---|---|---|---|---|---|---|
| Kew | 54 | 39 | 37 | 46 | 46 | 44 | 62 | 57 | 50 | 57 | 63 | 52 |
| Penzance | 127 | 89 | 81 | 64 | 62 | 50 | 70 | 75 | 80 | 111 | 124 | 123 |
| Birmingham | 76 | 55 | 48 | 55 | 63 | 47 | 68 | 69 | 60 | 70 | 78 | 68 |

*Mean temperature* (°C)

|  | J | F | M | A | M | J | Jy | A | S | O | N | D |
|---|---|---|---|---|---|---|---|---|---|---|---|---|
| Kew | 4·2 | 4·4 | 6·6 | 9·3 | 12·5 | 15·9 | 17·6 | 17·2 | 14·8 | 10·8 | 7·3 | 5·2 |
| Penzance | 7·2 | 6·8 | 8·3 | 9·9 | 12·2 | 14·9 | 16·4 | 16·6 | 15·1 | 12·4 | 9·7 | 8·0 |
| Birmingham | 3·2 | 3·7 | 5·8 | 8·7 | 11·7 | 15·0 | 16·7 | 16·4 | 13·9 | 10·0 | 6·6 | 4·4 |

## EXERCISE

Using the data given draw climate graphs for (*a*) Birmingham, (*b*) Kew Observatory, London and (*c*) Penzance, using a style similar to that in figure 2.1. Calculate the mean annual temperature, the range of temperature between the hottest and the coldest month, and the annual rainfall for each of the four stations. Find these stations in your atlas and describe the differences in climate.

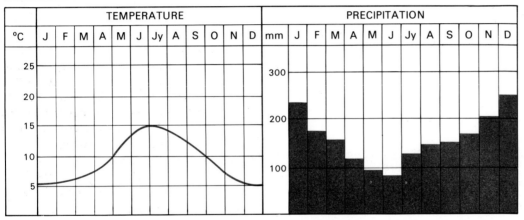

**Figure 2.1   Climate graphs for Fort William.**

The study you have made shows the great variation in climatic conditions from one part of Britain to another. One of the most significant components of climate is the rainfall. A rainfall map of Britain, when compared with a relief map, shows some correlation but on closer inspection it will be seen that it is the western uplands and particularly the western sides of these uplands which have most rainfall. Figure 2.2 illustrates rainfall totals experienced at various points across the Highlands of Scotland together with the altitude at these points.

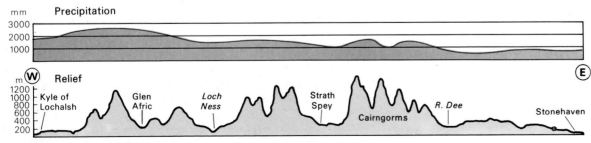

**Figure 2.2   Sections showing precipitation and relief across the Highlands of Scotland.**

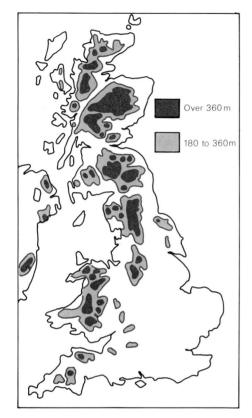

**Figure 2.3　Relief of Britain.**

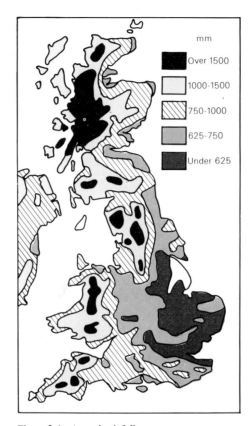

mm

■ Over 1500

□ 1000-1500

▨ 750-1000

▨ 625-750

■ Under 625

**Figure 2.4　Annual rainfall.**

'Climate' means average weather conditions. If you were to study the climate of the prairies of North America, for example, you might conclude that there is very little difference between climate and weather. For five months of the year mean daily temperatures are well below freezing and for only short periods of the day do temperatures rise above 0 °C. On the other hand the summers are characterised by long periods of clear skies and high temperatures broken by occasional heavy showers. This predictability is true of many parts of the world but we cannot describe Britain's climate in these terms. In fact a spell of one type of weather lasting for more than one week is remarkable. We do get such spells but more often the pattern of our weather is a changeable one with cloudy conditions, showers and sunny spells following each other throughout the year.

The British Isles has been described as being in the centre of a battleground between opposing air masses; between the polar air to the north and north-east, the sub-tropical to the south, the oceanic to the west and the continental to the east. First one is dominant, then another. The term 'air mass' is used to describe a huge volume of air which has similar temperature and moisture content throughout its mass at a given altitude. Such a mass will take on this uniformity when it remains stationary or moves very slowly over a portion of the surface of the earth where climatic conditions are themselves uniform.

The dominant rainbearing winds are those which come from the west and south-west. Relief rain occurs when moist air is forced to rise because high land is in its path. The air on rising expands and cools. Since cool air cannot hold as much water vapour as warmer air there is consequent condensation and precipitation. As the air stream crosses an upland area it is likely to become drier having lost moisture on rising; on descent the air is more able to hold the water vapour it contains as it becomes warmer.

(1) *Polar continental air:* The Eurasian land mass is a huge source region. In winter, cold, dry air can come in from Scandinavia or central Europe while in summer the air from this mass will be dry and warm.

(2) *Polar maritime air:* As the air mass moves southwards towards the British Isles it is warmed so that it can hold more water vapour. This air mass, which can influence the British climate particularly in winter, brings cold but slightly moister conditions than polar continental air.

(3) *Tropical maritime air:* The North Atlantic between latitudes 20 and 40˙ degrees north is occupied by a mass of stationary air which takes on a warm, very moist character.

(4) *Tropical continental air:* This air mass rarely reaches us in winter. It originates in North Africa and brings hot, dry conditions.

The air from these four source regions accounts for much of Britain's weather. It does not necessarily approach us along a direct path from the source region and the surface over which it passes will change the air's character. For example, in winter, polar continental air can come via Scandinavia directly across the North Sea, bringing cold, clear conditions with very little snow. If, however, it is diverted so that the airstream approaches Scotland from the north or even

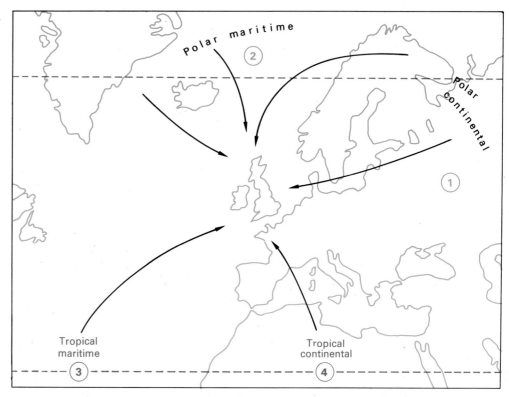

Figure 2.5 **Air masses influencing Britain's climate.**

north-west, more squally conditions with heavy snowfalls can be expected.

The dominant airflow is from the south-west and west, so the weather associated with this needs special attention. At any time during the year there are certain repeated patterns of weather and these are associated with the passage of low-pressure systems, known as depressions, across the country normally from west to east.

**An approaching warm front.** Identify the types of cloud on the photograph and state what changes in weather conditions might be expected in the next few hours. (*Photo: Patrick Bailey*)

41

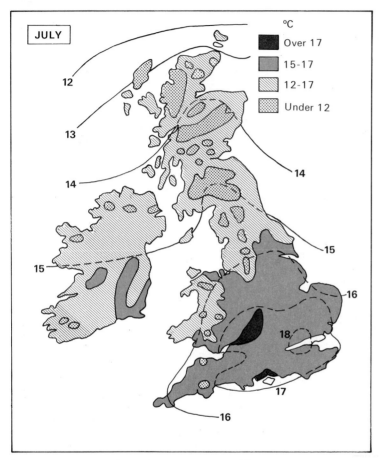
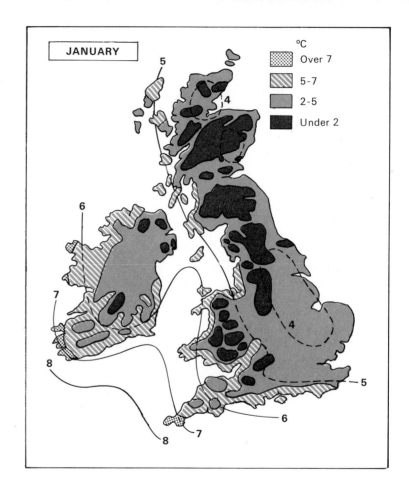

**Figure 2.6 Actual and sea level temperatures.** The shading shows the actual surface temperatures and the isotherms show temperatures reduced to sea level.

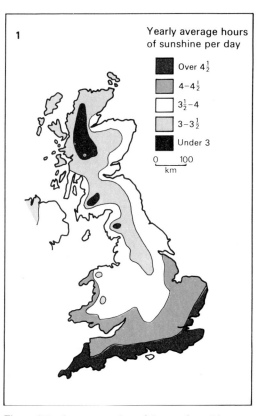

Figure 2.7 Average number of hours of sunshine per day.

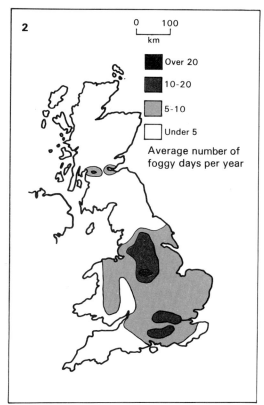

Figure 2.8 Average number of foggy days per year.

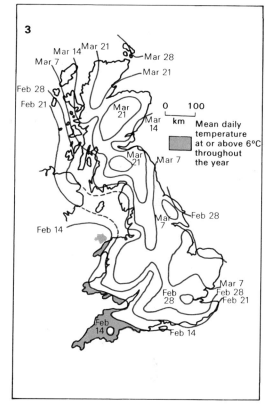

Figure 2.9 Dates when mean average daily temperature rises above 6 °C.

## EXERCISE

Temperature maps of large areas can give general conditions reduced to sea level or actual surface conditions. Study figure 2.6.

(1) Why is there a difference between actual surface temperatures and those reduced to sea level?

(2) What are the July and January temperatures for (a) Wick, (b) Glasgow, (c) Hull, (d) Bournemouth?

(3) (a) How does the isotherm pattern suggest that the sea has an equable influence on temperature? (b) Why do the isotherms have a north/south trend in winter and revert to an east/west pattern in summer?

Figure 2.7. Which groups of people should be particularly interested in the information on this map?

Figure 2.8. Where are the areas which experience the highest incidence of fog. Why is this?

Figure 2.9. The information on this map is of greater relevance to some types of farming than to others. Explain this.

**Hill fog.** Low-lying clouds in contact with hills are known as hill fog. (a) Why is this a common occurrence in upland areas of West Britain? (b) In what ways should it influence the clothing, equipment and precautions taken by hill walkers? (*Photo: Patrick Bailey*)

**Jet stream cirrus.** A jet stream is a ribbon of air moving rapidly at high altitudes just below the tropopause. The clouds associated with this are of the cirrus type, taking on a distinctive 'stepped' or 'ladder' pattern. The cumulus cloud on the photograph is at a much lower level. (*Photo: Patrick Bailey*)

# A DEPRESSION

There is a zone across the North Atlantic where tropical air and polar air come into contact. The meeting of the two contrasted air masses is known as the polar front. It is by no means static or regular in shape. Small waves form on the front and, as figure 2.10 shows, these may develop into depressions.

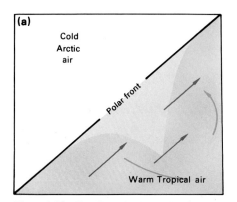
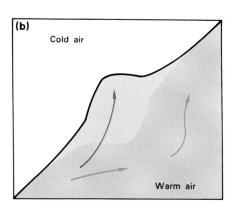
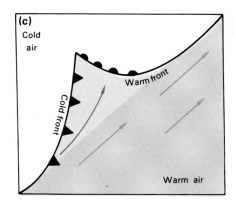

**Figure 2.10   The birth of a depression.**

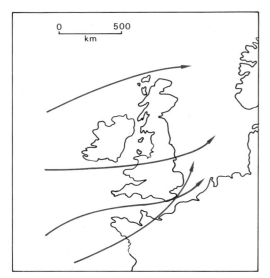

**Figure 2.11  Frequent courses of depressions.**

## EXERCISE

Describe the sequence of weather experienced over the course of two days at Birmingham as a depression approaches from the south-west and passes over the city. Base your description on the information given in figures 2.12 and 2.13.

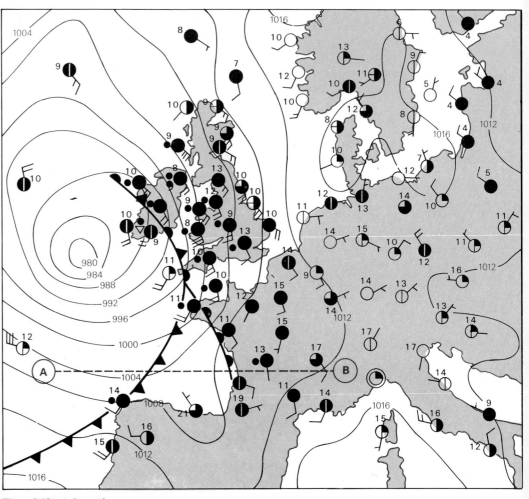

**Figure 2.12  A depression.**

46

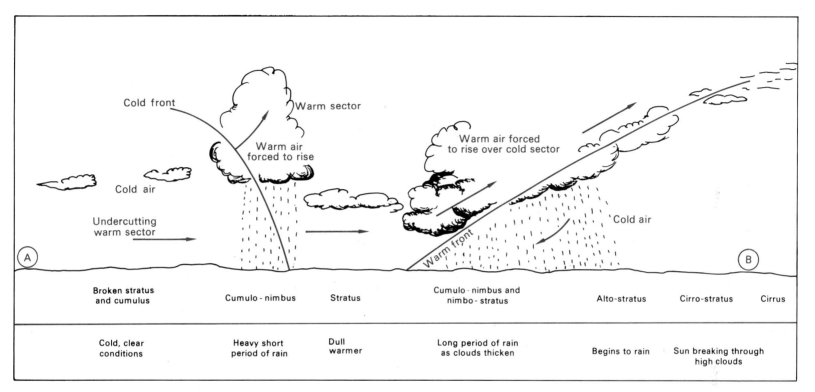

**Figure 2.13   Section across a depression along the line AB.**

# AN ANTICYCLONE

An anticyclone is a high-pressure system. In the centre the air is sinking, becoming warmer and spreading out. Notice how in figure 2.14 the winds blow gently in a clockwise direction around the centre of the anticyclone. Under these conditions the weather will be stable and dry, the winds light and the skies clear.

Map a in figure 2.15 illustrates a ridge of high pressure extending from the Netherlands in August. Such a system is likely to bring a hot, dry, sunny condition. A similar pattern, extending out from Denmark in January can result in a spell of bitterly cold, dry weather with daily temperatures below freezing.

A secondary depression may form in any part of the parent depression. Secondaries vary greatly in intensity but tend to be most developed when on the south side of a depression. Winds will be strongest in those

◀ **Temperature inversion at sunset.** Colder, denser air is in immediate contact with the ground and condensation has taken place; above this lies warmer air. This stable condition is the reverse of the normal when air temperatures decrease with altitude. (*Photo: Patrick Bailey*)

**Cumulus.** The cumulus cloud in unstable conditions can ▶ tower several kilometres above its base level. This indicates powerful upward movements of air and the possibility of rainfall. Why should this be so? (*Photo: Patrick Bailey*)

parts of the secondary furthest away from the parent depression. A frequent path for secondaries is up the English Channel from the Atlantic and it is the deep secondary similar to that illustrated in map b, which brings some of the most severe winds and heavy rain which southern England experiences.

Map c shows a col which is the central region between two highs and two lows. It is very difficult to forecast the weather for such a system but a col tends to be an area of light winds and clear skies. Fog may develop in winter. The system is unlikely to last for more than a day or two.

## EXERCISE

Consider the following weather conditions:
1. Gale force winds
2. Heavy, driving rain
3. Fog
4. Drifting snow
5. Freezing mist after rain
6. Black ice patches
7. Squalls and thunderstorms
  How do each of these affect (a) road (b) rail (c) air transport? How do we overcome these various weather conditions for each of the three stated forms of transport? How effective do you think our efforts are and how might they be improved?

## EXERCISE

List the differences between weather conditions over Birmingham on the maps of a depression (figure 2.12) and an anticyclone (figure 2.14) under the headings (a) Cloud cover, (b) Winds, their strength and direction, (c) Pressure pattern in the vicinity, (d) Rainfall.

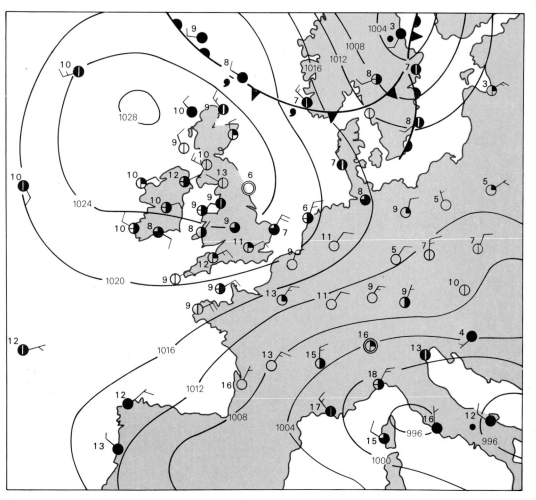

**Figure 2.14  An anticyclone.**

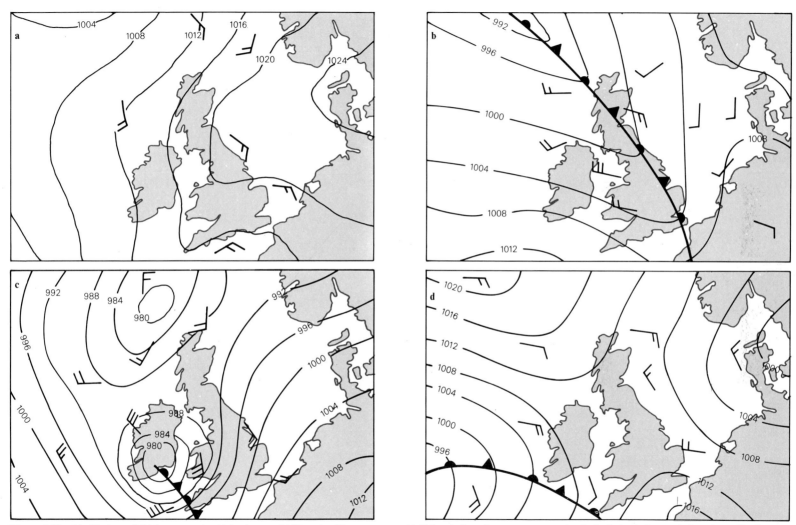

**Figure 2.15** (a) Wedge of high pressure which, in winter, brings dry, cold air from the continental interior.

(b) Trough of low pressure with light winds and a passing belt of rain along the zone of the occluded front.

(c) A secondary depression frequently brings violent weather including gale force winds and squally showers.

(d) A col which brings light variable winds and often bright weather.

**Hoar Frost.** Why does hoar frost occur?   (*Photo: Patrick Bailey*)

# Section 3
# FARMING, FORESTRY AND FISHING

# THE BACKGROUND

Today the general farming pattern in Britain is one of intensive mixed farming in which crops and livestock are both important and frequently closely integrated. For example, while stock fertilise the fields for arable crops the following year, many crops are grown to provide fodder for livestock. The farmer is aiming at as high a yield of produce from each hectare of his land as possible and to do this he will make extensive use of advanced scientific and mechanical techniques; he is increasingly occupied in deciding what fertilisers are required, what new machinery he must buy or hire, what rotation of crops to use and what feeding rota he must work out for the livestock.

However, there are many variations, not only from one region to another but within each region itself. The type of farming practised in any area is the response to physical, economic and social conditions. Physical conditions are concerned with the natural environment, particularly the climate, soil and relief; economic conditions are concerned with markets, profits, government help and costs of running the farm; social conditions are concerned with the way of life of the farmer and his family and in particular with the farming activities on which they prefer to concentrate. Social conditions also include the demands of society as a whole: we shall see how these have played a dominant part in the present pattern of farming.

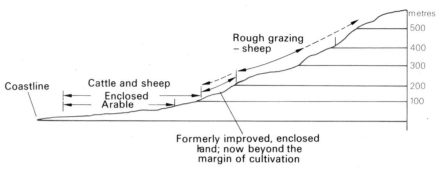

Figure 3.1  The influence of altitude on the land use of a portion of the North West Highlands of Scotland.

The farming response does not come automatically given a certain set of conditions but rather it comes as a result of many trials and failures over the centuries. Nor is the response, once a successful system has been found, necessarily permanent. The natural environment provides a range of possible farming activities. Recent years have seen remarkable and sudden changes in demand for meat, dairy and crop products. The farmer often has a difficult choice in deciding on what he should concentrate.

## Influences on the type of farming practised

The following section considers in more detail the various factors which influence the range of farming activities which are available to a farmer in a particular region.

## Physical factors

(1) *Relief.* As with so many of these influences that of relief can be best expressed on a sliding scale. Thus, the more rugged or the more elevated the relief becomes, the more limited are the choices available to a farmer. Of course, when we are considering elevation it is the climate experienced at different heights which is the most crucial feature. The degree of slope can, at a certain stage, make arable farming impossible because machinery cannot be used.

(2) *Soils.* The soil is an extremely complex mixture consisting mainly of (*a*) disintegrated rock particles which may be from underlying 'parent' rock or material transported from elsewhere and (*b*) decaying vegetation. As a soil develops it tends to form a series of distinctive layers or horizons. Often three such

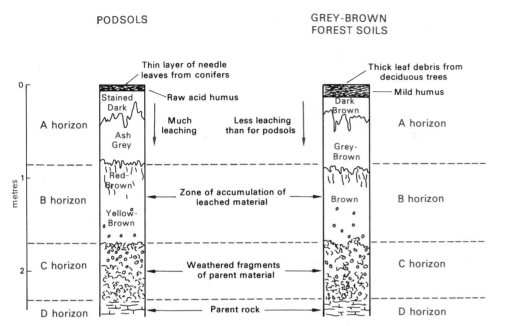

Figure 3.2 Soil profiles.

| Texture | Diameter of particles (mm) | Group |
|---|---|---|
| | Over 2 | Stones |
| Coarse material { | 0·2–2·0 | Coarse sand |
| | 0·02–0·2 | Fine sand |
| Intermediate | 0·002–0·02 | Silts |
| Fine | Under 0·002 | Clay |

Figure 3.3 The classification of soil particles.

horizons can be recognised. The upper, or 'A' horizon, has a large proportion of decaying vegetation so that there is a high humus content; below this the 'B' horizon is a layer with less humus and more rock particles and it is also a zone where minerals washed down from the 'A' horizon accumulate. Horizon 'C' is the weathered rock mantle.

Farming of the land breaks up the natural horizons. Not only deep ploughing and fertilising of the land but also improving the natural drainage and developing a certain crop rotation can alter a soil's character.

The *texture* of the soil means the size of the grains, which is of great significance to the farmer. An extremely sandy soil drains rapidly and lacks humus so that it is described as 'hungry'. The addition of farmyard manure can increase the humus content and the moisture-holding capacity temporarily, and by adding clay, a process called marling, the sandy soil can be permanently improved. At the other extreme is a clay soil which is made up of very fine particles: it holds very little air but does hold water which passes through with difficulty. A clay

soil is cold and farmers expect to wait later in the spring to work such soils, which are heavy to cultivate. Nevertheless clay soils can be made into very fertile agricultural land. The addition of lime not only improves the chemical balance (see below) but breaks up a clay to give a crumbly structure. Silty, alluvial soils are made up of particles intermediate in size between clays and sands and consequently tend to have an intermediate character. When adequately drained, alluvial soils can be very fertile.

The most valuable farmland is a loam.

(a) The pastoral farming scene is on Carboniferous limestone near Ingleborough. Explain why farming activities are limited in an area such as this. What types of farming are possible here? (b) The arable scene on the Cumbrian coast shows a dug-out portion of land with a thin mat of decaying vegetation which overlies a bed of peat up to five metres thick. Think of as many reasons as you can why this type of soil is so suitable for arable farming.  (*Photo: Farmers' Weekly*)

Such a soil is made up of particles of different sizes resulting in a crumbly character. A balanced loam is capable of retaining moisture, is well aerated and drains well; consequently it is readily cultivated. When sand particles dominate it is described as a sandy loam, and when clay particles dominate, a clayey loam.

The *chemical properties* of a soil are very important. Soils vary from very alkaline at one extreme to very acid at the other. When soil is subjected to much rainfall, or is constantly saturated, the bases are liable to be washed down to the 'B' horizon, a process called leaching. Consequently the soils become acid. This can be balanced by adding lime to the soil. Vital elements in the soil include nitrogen, phosphorus, calcium and potassium. Deficiencies can be made up by adding chemical fertilisers.

(a) Mountain soils. Normally thin. High degree of acidity. Vegetation—heather, coarse grasses, reeds, mosses. Use—rough grazing for sheep.

(b) Soils on gravels, sands and sandstones. Generally of poor fertility; much left as woodland and heath.

(c) Soils on chalk and limestone. Light soils, normally thin. Vegetation—a coarse grass. For long left as pasture. Now much used for extensive arable farming.

(d) Soils on clay vales. Heavy and acid. Often in need of drainage. Much permanent pasture.

(e) Soils on silts and drained lowland peats. Best arable land.

(3) *Climate*. A comparison of figures 2.4 and 3.4 does indicate a general association between rainfall total and the relative importance of arable and pastoral farming. The lowest rainfall totals occur in the east where transpiration of moisture from the soil is normally greatest and it is in these regions that arable farming dominates. Conversely rainfall is heavier in the west where grasslands

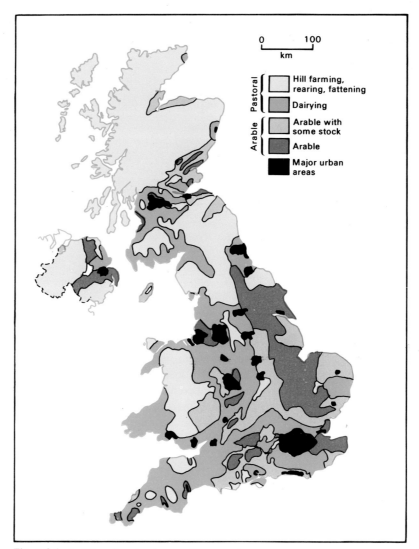

**Figure 3.4  Arable and pastoral areas.** The map shows the dominant farming system in a particular region.

are more widespread. Also note that areas with less rain are likely to be sunnier, conditions which are better for the ripening of crops, especially grain crops. Soil condition modifies this pattern locally. For example the heavy clays of Essex make it an area with a greater emphasis on pastoral farming while the rich peats of South Lancashire are devoted to market gardening and general arable farming.

A crucial temperature both for the growth of most crops and for soil bacteria to be active is 6 °C. The rate of growth is influenced by the amount of time during the day when the temperature is above 6 °C. Another significant feature for the farmer is that the soil just below the surface experiences more extreme temperatures than the air above.

| Air temperature | Soil temperature at 10 cm depth | Month |
|---|---|---|
| 4 °C | 2 °C | Jan. |
| 8 °C | 11 °C | April |
| 12 °C | 16 °C | May |
| 16 °C | 21 °C | Aug. |

The daily or diurnal range of temperature in the soil is greatly influenced by the cloud cover. When there is a clear sky mean temperatures rise during the day but fall markedly at night. On the other hand cloudy skies reduce the fall in soil temperatures at night. It follows that plant growth is greater

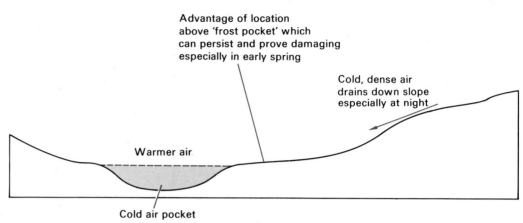

Figure 3.5 Frost pockets.

in mild, cloudy periods than clear, sunny ones when temperatures fall markedly at night. This factor is particularly important when considering spring weather conditions.

Besides rainfall and temperature there are other climatic factors to consider. The amount of sunshine is of significance not merely in ripening cereal crops but more generally in promoting plant growth. The average date of the last killing frost in spring and the danger of growing certain crops in frost pockets are other significant factors.

## Economic factors

During the twentieth century the rise in the standard of living of most people has led to increases in demand for certain farm products such as poultry, fruit, vegetables and flowers. At the same time many farm labourers have been attracted to better-paid jobs with shorter and regular hours of work. Consider the demands of a dairy herd, which must be milked twice a day seven days a week, or harvesting, when, if the weather is unsettled, weekend and evening work may be required to get the crop in. It is not surprising that the labour on so many farms is based on the farmer and his family: specialist gangs may be hired at harvest-time, and for making silage and hay.

The development of more and more specialised and complicated machinery in every branch of farming has made the job of the farmer lighter and eased the burden of finding labour. On the other hand the cost of the specialist equipment has led to many farmers concentrating on one aspect of farming

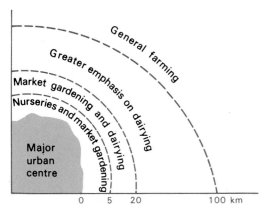

Figure 3.6 Concentric farming pattern showing the influence of a large urban centre.

when some years ago they might have been involved in several branches. The money spent on building and equipping a modern milking parlour is an obvious example: other examples ranging from equipment for intensive stock husbandry to that for arable and market gardening are dealt with in later sections.

The influence of the market shows some interesting features in the case of the farming pattern around large centres of population. There is a tendency for concentric zones of farming activity to develop, ranging from intensive methods closer to the market to more extensive systems further away. This zoning is frequently found to be the result of two major factors: (a) land values close to

urban areas are likely to be higher and competition from sources other than farming greater so that the farmer, to pay his way, must get as much return from each hectare of land as possible; (b) with many forms of intensive farming marketing of the produce may take place every day so that easy access to the market becomes a major factor. Compare for example the picking of greenhouse tomatoes with the harvesting of wheat: pickings every day at the height of the season on the one hand compared with one harvest a year on the other.

A final and increasingly important influence on farming these days is government policy. In recent years governments have given subsidies on many farm products including beef, milk, wheat and sugar beet to guarantee the farmer a certain price. This financial help to the farmer can encourage him to concentrate on a particular branch of farming. Nowadays this government influence is determined by the common agricultural policy of the European Economic Community. It is too early to assess the significance of this policy on the changing pattern of farming in Britain.

So far we have been considering influences on farming in general. Let us look at the national position of farming in Britain. Under 2 per cent of the United Kingdom's total labour force is directly engaged in farming and yet about three-quarters of the nation's land surface is given up to some form of agriculture. Over the past ten years

Britain itself produced on average 47 per cent of its total consumption of wheat, 30 per cent of its sugar, 70 per cent of its meat, 38 per cent of bacon, 8 per cent butter, 43 per cent cheese, all of its milk and nearly all the eggs and potatoes the country consumes.

In recent years, the trend in farming has been towards a greater yield of crops and more livestock on less land and with fewer workers. To achieve this farms have become bigger, the fields have been enlarged and there has been greater use of fertilisers and machinery. It is worth noting that fertilisers and machinery depend heavily on easily available sources of energy. As these begin to peter out we may, looking ahead several generations, have to face drastic changes in our farming habits.

# HILL FARMING

About one-third of Britain's farmland is classified as upland or hill-farming country. Much of this category is characterised by shallow, impoverished soils which are frequently very acid. Coupled with this, over much of these areas a bleak and harsh climate is experienced so that the vegetation response consists of heather, bracken and coarse grasses. The keeping of upland sheep which have been bred to exist in such an environment is likely to be the only possible agricultural activity over large tracts of the region.

For long this has been the home of the subsistence farmer who could just exist on the farming economy he had developed but made very little profit, providing most of the needs for his family from the farm holding itself. The focus of this farming economy has been based on the lower valleys or coastal strips. Here small, enclosed fields provide winter fodder (oats and swedes) and grazing for the sheep as well as vegetables for the farmer. Here also are the farm buildings and the pens for lambs. On the hill-sides the natural grasses and heather provide late spring, summer and autumn grazing for the sheep. In some areas sheep are kept on lower hill-sides throughout the year. For long the factor limiting the expansion of the holding and the number of sheep kept has been the availability of winter fodder.

On holdings of 800 hectares or more with normally 95 per cent of the land rough grazing, the aim is to produce lambs as well as wool from ewes. During this century there have been large numbers of lambs and ewes sold to lowland farms for fattening or for breeding purposes. Recently there has been a tendency for more farmers to improve pasture on lower slopes and valley floors and to fatten the lambs themselves. This trend is partly due to a move by lowland farmers away from fattening lambs because of the relatively low profit as compared with alternative farming activities.

In hill-farming areas there has been an ex-

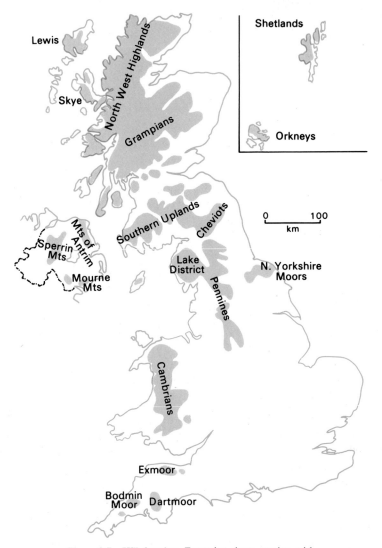

**Figure 3.7  Hill farming.** Extensive sheep rearing with some cattle.

**Teesdale.** What do you understand by the term 'hill farming'? (*Photo: Patrick Bailey*)

pansion of other activities in recent years, reducing the amount of land taken up with sheep farming. There are cattle rearing farms and, in the lower areas, dairying as well. Hydro-electric schemes, military training grounds and land for afforestation have all meant the appropriation of large sections previously devoted to hill farming while land for recreational facilities is now in much greater demand than ever before. The main trends then are that lowland sheep farming is declining and upland sheep farming is likewise being subjected to pressure from more profitable activities.

There may in the future be improved opportunities for hill farming. For example there are plant-breeding experiments taking place by which new strains of coarse grasses have been sown on high land. These have nitrogen-fixing properties which improve the fertility of the ground and consequently increase the number of sheep which can be grazed per hectare. There are also experimental schemes in which Highland Deer are being bred as a food source.

# CROFTING

Crofting is a specialised form of hill farming. The term is used to describe the pattern of farming existing in parts of North West Scotland. A croft is a small rented piece of land, usually ranging from 5 to 25 hectares. Crofts normally have developed in groups or 'townships' along coastal lowlands both on the mainland and islands. The basis of the crofting economy is sheep: for much of the year these may be kept on common grazing adjoining the croft. On the holding, fodder crops include oats, barley and turnips as well as hay. Beef cattle, which are sheltered under cover in the winter, and pigs are sometimes kept.

This has been the traditional system in much of the Highlands. During the twentieth century crofting areas have experienced considerable depopulation as the younger mem-

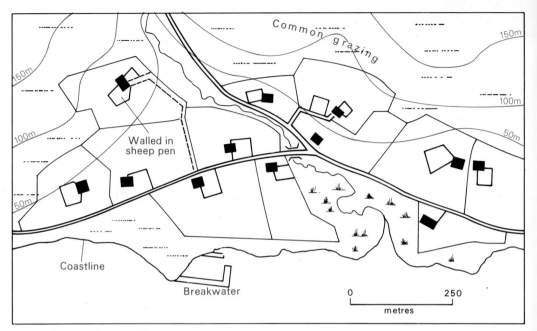

**Figure 3.8  A crofting township.**

bers of the family were attracted away to the more comfortable living available in the cities. Abandoned, decaying crofts are a familiar site along many parts of the west coast of Scotland. The crofters who remain have often adapted to change. For long crofters had supplemented their living by fishing or handicrafts. Now many, living in bigger and more comfortable crofts than their predecessors, provide bed and breakfast for tourists or have extended croft industries, such as weaving and wood carving, aimed at the tourist market.

# BEEF CATTLE PRODUCTION

Most beef is obtained from cattle slaughtered when ten months to three years old. In recent years only 30 per cent of British beef has come from beef herds such as Hereford, Aberdeen Angus, Galloway and Shorthorn. This beef comes either from farms where the cattle are bred, reared and fattened or from farmers who bought in young cattle and fattened them. About 70 per cent of the beef is from dairy herds such as Ayrshire, Friesian and Guernsey; these may be slaughtered dairy cows or calves taken from dairy herds and fattened.

## Stock rearing

Cattle-rearing farms lie mainly on the margins of the hills in areas which have a more favourable climate and richer soils than the hill sheep farms. Such holdings are smaller than hill sheep farms and have a lower proportion of rough grazing and permanent grass. A typical holding will have about half its land under rough grazing and on the arable section oats and hay will be the main crops. On such a holding as is illustrated there are unlikely to be any hired workers since labour demands are relatively light. Young stock receive very little attention until weaned. At that stage the beasts may be kept on the farm or sold. Such cattle will probably be store cattle; that is they will not be fattened but grazed out-of-doors. However, supplementary feed-stuffs—hay, barley and roots—will normally be provided; these are grown on the farm.

When between one and two years old beef cattle are fattened. This may well take place on the same farm: such dual-function rearing and feeding farms are generally situated on richer land than farms that are only concerned with rearing. Alternatively the store cattle are sold to a farm devoted to the fattening of beef cattle.

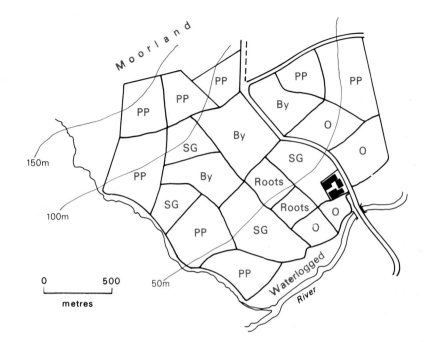

**Figure 3.9 A cattle rearing and fattening farm, Northumberland.**
The size of the holding is 274 hectares. What is the average size of each field? What do each of the abbreviations stand for? In what ways does relief influence the field usage?

## Cattle fattening

A cattle-fattening farm is likely to have a considerable proportion of land under grass (what percentage on the example given here?), this being used not only for grazing but also for hay and silage. Arable land produces turnips, mangolds and kale as winter feedstuffs, to which prepared concentrates are added. A farmer may buy in stock both in autumn and spring: in this case he will work on a six-month fattening rota during which time it is hoped that the cattle will put on about 100 kg in weight (from between 350 and 450 kg to between 450 and 550 kg).

Revolutionary changes have taken place in recent years towards more intensive systems of fattening using 'controlled environment housing'.

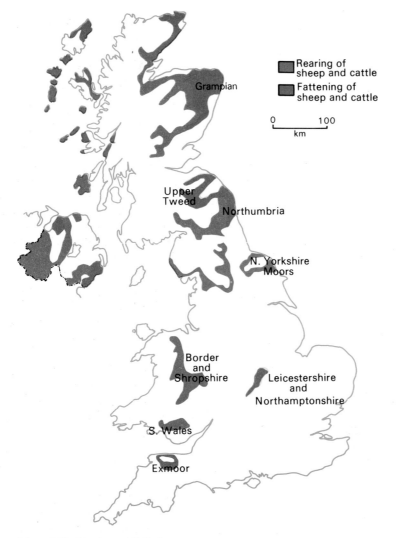

**Figure 3.10  Rearing and fattening.**

Map labels:
- Rearing of sheep and cattle
- Fattening of sheep and cattle
- 0 — 100 km
- Grampian
- Upper Tweed
- Northumbria
- N. Yorkshire Moors
- Border and Shropshire
- Leicestershire and Northamptonshire
- S. Wales
- Exmoor

# DAIRYING

Dairy farming is very scattered over the lowlands of Britain although the greatest concentrations are in the western lowlands of England. Areas with a rainfall of over 750 mm a year and with soils which range from medium to heavy, so that they are capable of holding moisture, support a rich grass. It is these areas which have a greater dairying concentration although the influence of nearby large markets is often a factor.

Although some dairy farmers combine milk production with mixed arable farming and others with the raising of store cattle, the largest proportion concentrate on the dairy herd as their principal enterprise. They tend either to practise pasture dairying or dairying with some arable. The former holdings are concentrated on the heavy soils where a high rainfall produces a good grass but the land is more difficult to cultivate than on light loams. The Cheshire Plains and Somerset are examples where this form of dairying dominates. On the other hand over much of the Midlands, the London and Hampshire basins and in Scotland and Northern Ireland, fields are ploughed periodically to grow sugar beet or fodder crops such as oats, barley and kale. This is a complementary system, the cattle fertilising the fields later destined for arable crops, and these crops in turn providing feedstuffs for the cattle: for example the sugar beet provides tops and pulp for cattle.

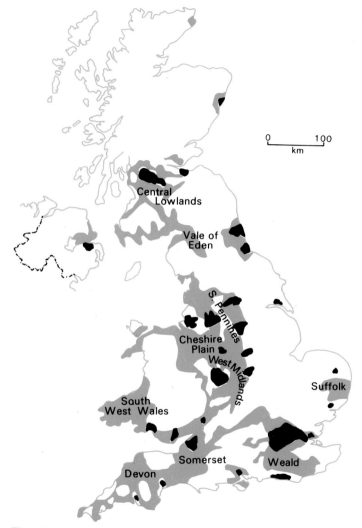

**Figure 3.11  Areas where dairying is a dominant activity.**
The main urban areas are shown in black. Why have
they been marked in?

Central
Lowlands

Vale of
Eden

S. Pennines

Cheshire
Plain

West Midlands

Suffolk

South
West Wales

Somerset

Weald

Devon

0        100
km

## Account of a Dairy Farm located in the Midlands

The farm illustrated in figure 3.12 is a 40
hectare holding entirely under grass. It is
located on undulating land, the bedrock
being Keuper Marl with shallow cappings of
boulder clay in places and some alluvium;
the soil varies from medium to heavy.

The holding has a herd of pedigree British
Friesians. On average fifty-five cows are in
milk at any time and about sixty calves are
born each year. The aim is for the cows to
calve once a year. Remember that cows do
not produce milk unless they have calves.
The heifers (young cows that have not
calved) are retained by the farmer so that
they can be introduced into the dairy herd
and the bulls are sold.

During any one year 12 hectares are
grazed by the dairy herd which is moved
around every three weeks from one paddock
to another. The young stock graze 8 hectares
of land and the remaining 20 hectares are
kept for silage. This grass is cut green twice
a year in late May or early June and in early
August, compacted into a heap and covered
so that it ferments. Seven tonnes of silage is
needed to feed one cow through the winter.

All except the very young calves are
grazed from about April to late November.
During this time concentrates are given as a
supplement both to those heifers which are
about to calve and to cows in milk. During
the winter months, when stock are housed

65

under cover, the cows have 36 kg silage and some rolled barley or sugar-beet pulp each day.

The amount of milk produced varies during the year but a cow can produce between 3000 and 7000 litres of milk annually, the daily production from the herd varying from 500 to 900 litres.

The farmer works single handed, milking in a six-berth shed twice each day of the year from 6 until 8.00 a.m. and 4 until 6.00 p.m. Contractors are employed for silage making and bale carting. Ten tonnes of hay and a hundred tonnes of straw are bought in.

Each year about 25 tonnes of nitrogen fertilisers are applied to the grassland and occasionally phosphates, potash and lime are also added. Of course cow dung helps to keep the grasslands in good condition.

## PIG FARMING

Most pigs are kept on mixed farms. There tends to be a greater concentration of pigs near large centres of population, such as in the West Riding of Yorkshire and South East Lancashire, and in dairying areas, such as Cheshire and East Cornwall. The latter concentration is probably an historical one; at one time the skimmed milk and whey from the farm dairy provided an important part of the pigs' diet but this is no longer available; however, the traditional interest in pig farming remains.

In recent years pigs have been cross-bred to obtain a beast which is quick growing and with not too much fat. At the same time successful cross-breeding has also provided pig-farmers with pigs which have larger litters. As with beef cattle production there are holdings which both breed and fatten pigs while others specialise in one or the other. A holding which concentrates on fattening buys in the piglets when they are six to eight weeks old and weigh under 20 kg. They are fattened for perhaps four months, at the end of which time, when weighing about 100 kg, they are ready for the bacon factory. There is a move towards more intensive farming methods particularly on the specialist pig units. On some of these the pigs are fattened indoors under a controlled environment system with both feeding and cleaning done automatically.

## POULTRY FARMING

This particular aspect of farming has expanded rapidly in the last quarter of a century as poultry has become a relatively cheaper meat. Units of production are often large both for egg production and in the broiler industry. In both, mass production is

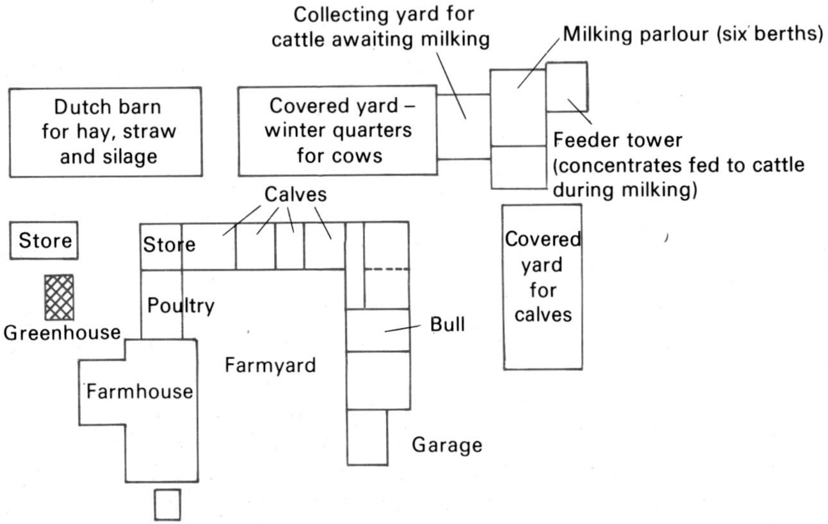

**Figure 3.12  Buildings on the dairy farm.**

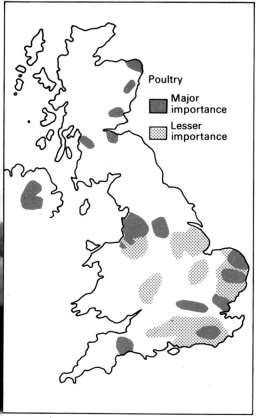

Figure 3.13 (a) Poultry and (b) Pigs.

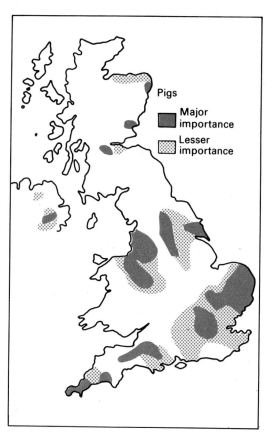

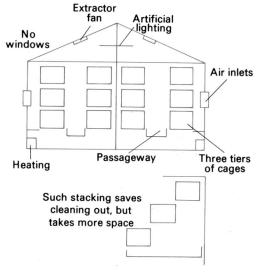

Multiple cages taking 25 birds each.
Such a house could hold 10 000 broilers

Figure 3.14   A controlled environment poultry unit.

now commonplace, and occurs in an artificial environment where temperature, light, air circulation, relative humidity and feeding are automatically controlled.

These conditions, for broilers, are obtainable in the 'deep litter' houses where the broilers live on a layer of moss, shavings and straw. Here food is available always and a dull light is provided for most of the time. In this way broilers are ready when nine to ten weeks old. The large poultry units have their own processing plant where plucking, cleaning, packing and freezing are all done. Such a plant has perhaps 50 000 broilers at a time. The intensive units for egg production consist generally of long sheds where tiers of cages are separated by narrow passageways.

Poultry farming, particularly artificial environment units, is little influenced by natural conditions so that distribution is very widespread, extending from the Orkneys to Cornwall. However, there is a greater concentration around urban areas in South East England, Yorkshire, Northern Ireland and Lancashire.

# ARABLE FARMING

An arable farm is normally reckoned to be a holding with at least two-thirds of the land under crops and of this cropland less than one-quarter is normally for fodder. The distribution of areas where much of the agriculture is devoted to arable farming is shown on the map (figure 3.15).

### EXERCISE

From evidence on the map, what can you say about the location of areas where arable farming dominates? What the map does not show is that most of the areas noted for arable farming have light, easily workable loam soils.

When considering the list of arable crops there should be a distinction between crops which are sold and those grown to feed livestock. The former are known as cash crops, the latter as fodder crops. Of the cash crops grown in Britain, wheat, sugar beet and potatoes are the most important and very often these are grown in rotation on the same farm. Although all three crops will tolerate a wide range of soils, optimum conditions are deep, well-drained loams rich in humus which gives a high moisture retaining capacity so the soils do not dry out quickly.

Successful arable farming relies very much on mechanised techniques. Mechanised planting and harvesting of cereals is universal in this country. Sugar beet similarly is planted and harvested by mechanical means

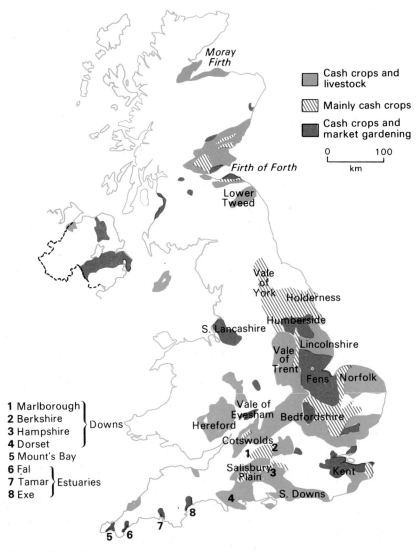

Figure 3.15 Arable areas.

**The Fens.** Describe the scene commenting on the absence of hedges, the raised road and the colour of the soil. (*Photo: Patrick Bailey*)

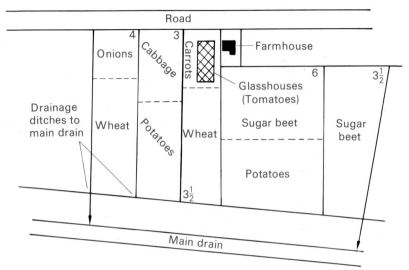

Labels within figure:
Road
4
Onions
3
Cabbage
Carrots
Farmhouse
Drainage ditches to main drain
Wheat
Potatoes
Wheat
Glasshouses (Tomatoes)
Sugar beet
6
Sugar beet
$3\frac{1}{2}$
Potatoes
$3\frac{1}{2}$
Main drain

**Figure 3.16  Arable farms of 20 hectares near Spalding in the Fens.**

but there is still a lot of hand hoeing required. Potato cultivation makes much bigger demands on labour. Harvesting in particular, although it involves the crop being dug out by tractor, also requires it to be picked up by hand. Sugar beet is considered a 'safe crop' by farmers since they can obtain a market guaranteed at a given price by the British Sugar Corporation.

'Field vegetables' is the term given to a group of vegetables such as peas, carrots, onions, leeks, parsnips and celery when they are grown as cash crops on arable farms. As we shall see later they are widely grown on market gardens also.

Barley occupies more land in this country than any other cereal while the area under oats has decreased in recent years. Oats will tolerate a much wider range of physical conditions than barley. It will ripen with much less sunshine and will tolerate far wetter conditions and poorer soils. The cultivation of barley has expanded particularly on the chalklands. The oats crop provides grain and straw for livestock; barley has this use as well but it is needed also for brewing and distilling.

Kale, cabbage, rape, mangolds, swedes and turnips are the other main fodder crops along with the two fodder cereals, barley and oats. Most are used on the farm, being either grazed where they grow or harvested and stored until

required. There has been an overall decline in fodder crops. The greater use of silage made from grass, sugar beet tops and manufactured feedstuffs all help to account for this.

Figure 3.16 is the plan of an arable farm of 20 hectares worked by family labour—farmer, wife and son. As with so many of the holdings on the light silt soils near Spalding, the farmer grows not only three basic farm cash crops, wheat, sugar beet and potatoes but also a variety of market garden crops, carrots, cabbage and onions as well as tomatoes grown under glass and sold locally.

## The farmer's year—based on the arable farm study

| | |
|---|---|
| *January and February* | Maintenance work about the farm. Clean out ditches and drains. |
| *March* | Fertilise sugar beet and potato fields. Sow carrots. |
| *April* | Early on, sow sugar beet in drills pulled by tractor. Plant out cabbages. Towards the end of the month plant potatoes (mechanised planter) and thin out sugar beet seedlings by hand hoe. |
| *May* | Complete hoeing sugar beet. Spray wheat to kill weeds. |
| *June* | Spray potatoes as a protection against blight. Start cutting cabbages. |

| | |
|---|---|
| *July* | Lift early potatoes. Continue cutting cabbages. |
| *August* | Begin digging carrots—this can continue periodically into the next year. Sometime between the end of August and mid-September the combine-harvester comes in with its crew to harvest wheat. |
| *September* | Burn stubble. Lift main-crop potatoes. Plough wheat fields. Start harvesting sugar beet: this can continue until December. The beet harvester used will top, clean and load the beet on the trailer. |
| *October* | Sow winter wheat. |
| *November and December* | Plough again. |

## EXERCISE

List the jobs which are performed (*a*) by machinery and (*b*) by hand. What machinery is used? Can you assess how much it cost? To what extent is the holding mechanised? How could it be more mechanised?

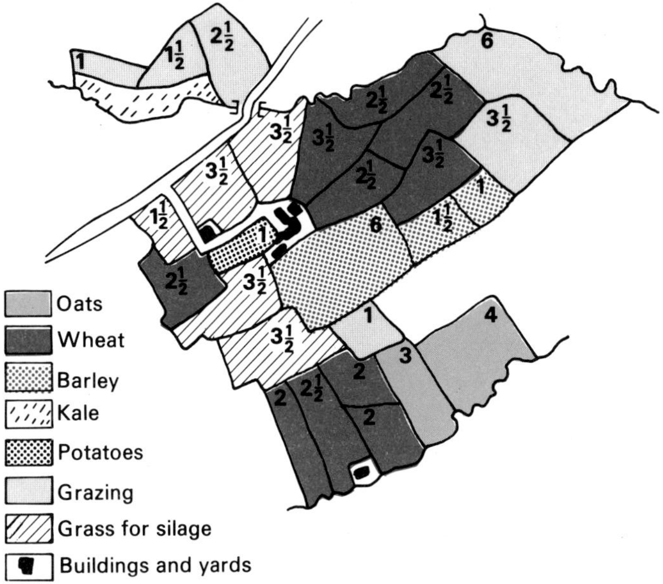

**Figure 3.17  Plan of a mixed farm.**

This is a 75 hectare holding which has grazing rights over 28 hectares of nearby parkland. The soils are generally heavy consisting of marls overlain with uneven deposits of glacial drift consisting of boulder clay, sands and gravels. The farmer works the holding with the help of two sons and occasional casual labour. There are 35 Friesian cows in milk. 45 young stock being fattened for beef, 120 ewes (for lambing and wool) and 190 lambs (for meat). The holding has two tractors, a plough, seed drill, crop spray, combine harvester, Cambridge roll and chain harrow.

What advantages and what disadvantages can you see in this system compared with methods of farming which concentrate on one activity?

# MIXED FARMING

Mixed farming is a system involving both the production of cash crops and keeping of livestock, but there is a wide range of variations. Do not confuse these farms with those which are based on livestock and where the arable land is used exclusively for providing feedstuffs and bedding for the livestock kept.

The advantage of mixed farming is that the farmer spreads his risks and with rapidly changing costs and prices can switch the emphasis of his attention from one branch of farming to another. On the other hand a disadvantage may be that the farmer, because of the wide range of his activities, may not be able to afford the price of a piece of equipment which a specialist farmer would consider essential.

**Figure 3.18 Major vegetable producing areas.**

# MARKET GARDENING

This is a very intensive form of farming concerned with the production of vegetables, salad crops and flowers. For long most market gardens have been concerned with supplying produce daily to a market in a large town over as long a season as possible. Recently more and more market gardeners send the whole of a particular crop to a processing plant.

**EXERCISE**

Figure 3.18 shows the distribution of market gardening areas. Some of the influences on the location of these areas are listed below. Which of these influences do you think is the strongest for each of the areas shown on the map?

(1) The best soil is that which is light, easy to work, well drained and will warm up quickly in spring. Thus a sandy loam is better than a clay.

(2) An early spring will give a region an advantage for it will be able to cash in on an early market when prices are high. The amount of sunlight is another climatic factor.

(3) Since crops can be marketed every day for many months, proximity to market is an advantage. Similarly many perishable vegetable crops deteriorate rapidly on long journeys.

When you complete the exercise you will probably conclude that no area fits the ideal conditions on all three counts.

Of all the market garden crops, peas are the most important. Only 20 per cent of this crop is sent to the market as a fresh crop. The remaining 80 per cent goes to processing plants and half of this is frozen and the rest canned or dried. Much of the crop is grown under contract whereby the whole crop goes to the processer. Peas for processing are picked and shelled by a pea viner and then sent immediately to a processing plant before the peas deteriorate. The large freezing firms consider that all peas should be frozen within ninety minutes of being picked. Within what distance from the plant do you think the peas should be grown—150 km, 50 km, or 25 km? Allow for collection, loading, unloading and shifting to the freezer.

Bedfordshire has one of the major concentrations of market gardens. Here on sandy soils and river gravels the main crops are cauliflowers, cabbage and brussels sprouts together with chrysanthemums, while tomatoes and lettuce are grown under glass. Markets for this region range from the north of England and the Midlands to the London area. In an attempt to reduce demands on labour, especially during the harvesting periods, more and more mechanisation is being introduced, particularly on the larger holdings which can afford it. The brussels sprouts combine harvester is a good example of this.

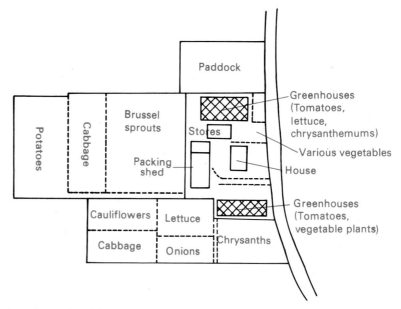

**Figure 3.19  A market garden of 8 hectares.** A farmer and his wife employ three people permanently and there is some casual and part-time help at busy periods during the year.

## The year's work on the Bedfordshire holding

| January | Sow lettuce and cabbage seeds under glass. Take chrysanthemum cuttings. Pick brussels sprouts. |
| February | Continue picking brussels sprouts and sowing brussels sprouts and lettuce seeds under glass. |
| March | Plant out lettuce, cauliflower and cabbage seedlings. |
| April | Continue this planting out as well as planting of potatoes and chrysanthemums. |
| May | Plant out brussels sprouts. Spray chrysanthemums |
| June July | } Cut cabbage and cauliflower. |
| August September | } Cut lettuce; dig potatoes and pick chrysanthemums. |
| October | Pick chrysanthemums. |
| November December | } Plough; pick brussels sprouts. |

73

# GLASSHOUSE CULTIVATION

This is the most intensive form of crop production. Many market gardens have glasshouses which may be used to provide plants for later planting out as well as crops grown solely under glass. However, some areas of the country have concentrations of specialist glasshouse holdings: amongst these are the districts east of Blackpool and portions of the Fens, Sussex and Kent.

Originally glasshouse cultivation developed near to large urban centres, the Lea and Clyde valleys being examples, but such areas no longer have such a favourable location and many glasshouse holdings have recently disappeared, particularly from the Lea Valley. Pollution and difficulties in obtaining labour (which has been attracted away to better-paid jobs) are factors. With improved transport during the middle of the twentieth century the distribution of commercial glasshouse holdings has become more scattered and new holdings are more likely to be set up in sunnier districts, for example along the south coast of England.

The main crops grown in heated glasshouses include tomatoes, cucumbers, carnations and chrysanthemums. Yet again there are trends towards more and more automatic methods of heating, watering and even feeding of glasshouse beds.

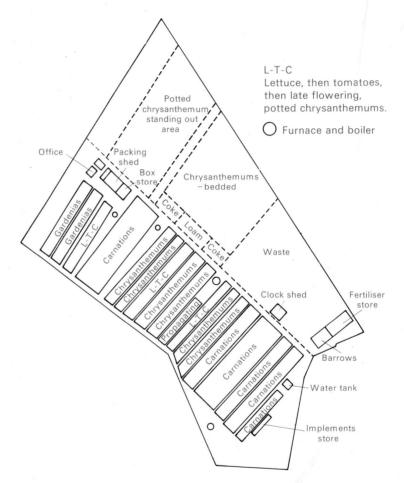

**Figure 3.20   A nursery garden.**

# FRUIT FARMING

There are two distinct forms of fruit farming: one group produces orchard or top fruit such as apples, pears and plums, and the other cultivates small or soft fruit such as strawberries, raspberries and blackcurrants.

The site of an orchard should avoid areas where frost pockets are liable to form, especially in spring when the blossom is set. It should also avoid areas exposed to strong winds: these can be particularly damaging when the fruit is nearly ripe. Therefore the best sites for orchards are likely to be sheltered, sloping sites.

The main areas noted for small fruit farming tend to have higher average sunshine figures and lower rainfall totals than the orchard areas but they also favour sloping land.

## EXERCISE

Comment on the distribution of fruit-farming areas as shown in figure 3.21.

Some of the activities on a fruit farm make particularly heavy demands on labour. Both spraying and pruning are crucial and time-consuming jobs but it is the harvesting of the fruit which gives the farmer most headaches and requires the employment of much casual labour. As with market gardeners many fruit farmers have contracts with processing firms such as canning, freezing and jam-making concerns.

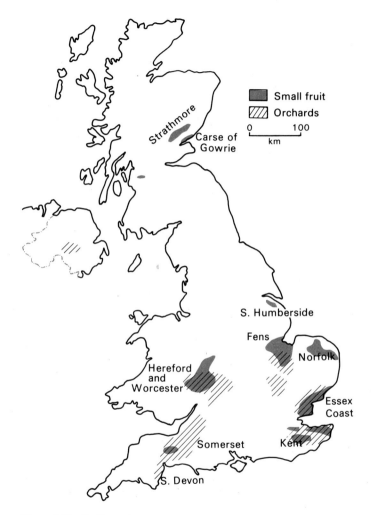

Figure 3.21 Fruit growing.

The scenes show strip grazing, potato harvesting and preparing the soil in a multiple-span glasshouse. **1.** What do you understand by the term 'strip-grazing' and what are its advantages? **2.** Mention different methods used by farmers to overcome the problem of finding additional seasonal labour particularly during harvest time. **3.** (a) What do you understand by the term 'intensive cultivation'? (b) List the cost items involved in glasshouse cultivation. (c) What types of crops are grown under glass? (*Photos: Farmers' Weekly*)

## FORESTRY

When the Romans left Britain much of the land, and in particular the clay lowland, was still thickly forested, but by Stuart times there remained only scattered remnants of this natural cover of woodland. The timber had been cut to be used for fuel or for building material and the ever-expanding clearings could be used for farming. In the eighteenth and nineteenth centuries there began a period of conservation and reafforestation. Landowners planted belts of trees on a large scale both for economic reasons, to replace dwindling resources, and for aesthetic reasons, to beautify the landscape of their estates.

From the 1920s private efforts have been largely superseded by a national policy of both afforestation and reafforestation undertaken by the Forestry Commission. The Commission has taken over considerable areas of land which is agriculturally of low fertility. This includes not only stretches of moorland in the upland areas of Britain but also many sandy lowland areas.

Seedlings are raised in nursery beds for a period of from two to four years and then they are planted out in blocks. Most of the forestry schemes are coniferous plantations and the risk of fire is greater for these than deciduous trees. For this reason there are a series of regulations concerning the fire lanes which separate the blocks. These lanes also act as access roads. When the young trees

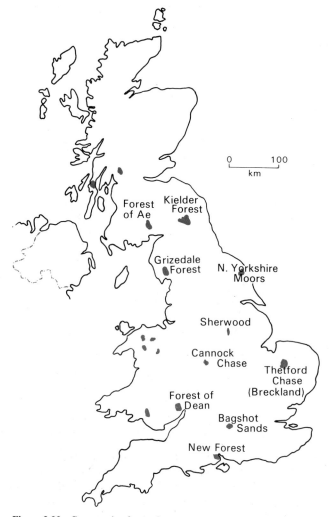

**Figure 3.22   Some major forested areas.**

are set out it is frequently necessary on wetter upland slopes to dig drainage ditches and also provide an initial application of fertilisers. Thinning out begins after about twenty years and continues every five to ten years until the cycle of eighty years or more is completed. Thinnings can be used for chipboard, pulp, paper and fencing; mature trees can be used for constructional timbers, such as planks and beams, for the building industry.

Various species of conifer are planted. In lowland areas Corsican and Scots pine are favoured since they are sufficiently hardy to flourish on poor sandy soils. Deciduous trees are often planted around the edge of individual conifer blocks to improve the appearance as well as to assist as fire breaks. In exposed upland areas, particularly on ill-drained acid soils, the Sitka spruce responds better than most other conifers.

# FISHING

The waters around the shores of the British Isles have provided the British market with a great variety of fish. Many species of fish considered edible are found in shallow water and the continental shelf off north-west Europe provides such an area. In particular the North Sea, which is mostly less than 100 metres deep, has been a major fishing ground. There are considerable variations in temperature in British coastal waters due to

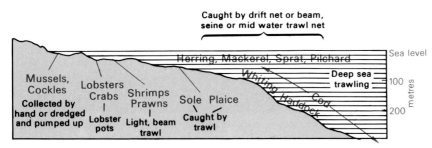

**Figure 3.23   To show the influence of depth of water on location of different species of fish.**

the mixing of ocean currents and this factor also helps to account for the variety of fish. The North Atlantic Drift is especially significant since it brings valuable mineral salts, the food for plankton, which are the organisms which fish feed on.

It is not surprising, with such an enormous market in the densely populated countries bordering the North Sea, that overfishing in the nineteenth and first half of the twentieth century in local waters has depleted shoals drastically so that fishermen are now forced to go further and further away in search of new fishing grounds. Consequently the number of smaller vessels which fish the shallow waters of the North Sea have been decimated. This cut-back has been particularly marked in recent years: for example Lowestoft's fishing fleet consisted of 120 vessels in 1973, 80 in 1974 and 56 in 1975.

International agreements made in the mid-seventies, designed to protect national fishing grounds, must lead to changes in the

**Figure 3.24   Fishing ports and the continental shelf.**

pattern of the British fishing industry. A reduction in the deep sea fleet, which, in the past, has operated off the coasts of northern Norway, south west Greenland, Newfoundland and Iceland as well as in the Barents Sea, seems inevitable.

Coastal waters provide the fishing grounds for pelagic fish in particular. This group of fish includes herring, pilchards, mackerel and sprats. Pelagic fish for long have been caught mostly from drifters using a series of drift nets which may stretch for several kilometres. The nets hang vertically in the water, held up by a line of buoys, and drift

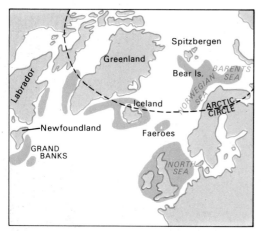

**Figure 3.25  Fishing grounds used by the British fishing fleet.**

with the wind and tide: the fish swim into the nets and are caught by their gills. The most valuable catch is herring which, like other pelagic fish, feed near the surface at night. The herring form shoals covering an area of perhaps 5 kilometres across and 10 kilometres long. For centuries shoals have concentrated off the Orkneys in June, off the Yorkshire coast in July and August and off the East Anglian coast in October. Overfishing has so depleted the stock of North Sea herring that in recent years the build up of a shoal is very unpredictable. Today the most valuable herring catches are made off the north-west coast of Scotland by trawling and not by the traditional drifting methods.

Demersal fish live on or near the bottom of the sea and include cod and haddock as well as whiting, hake, sole, plaice, halibut and turbot. Until recently about three-quarters of the landings at British ports were of demersal fish and of these 55 per cent were cod and about 25 per cent haddock. Demersal fish are caught by trawling. A trawl is a bag-shaped net open at one end which is towed slowly along the sea bed by a trawler. In shallow water around the coasts of Scotland vessels use the seine net which is similar to a trawl net but with long wings of netting leading to the net bag; a warp or tow line is attached to each wing of the net and one end is secured to a marker buoy; the vessel then sails on a triangular course paying out the warp and 'shooting' the net. When it returns to the buoy the net is winched in.

Much of the demersal catch is brought in by the fleets of deep sea trawlers which are based on Hull and Grimsby and which operate off the coasts of northern Norway, south-west Greenland, Iceland, the Barents Sea and sometimes even as far afield as Newfoundland. The deep sea trawlers are equipped with freezer holds and in many cases with facilities for filleting fish and some even prepare fish meal and oil from the waste while at sea. In contrast to inshore vessels, which are largely owned and operated by the fishermen themselves, the deep sea vessels are owned by companies.

Another aspect of fishing is the catching or collecting of shellfish. Shrimping is especially important in the Wash and Morecambe Bay, lobsters are caught off the Channel Island coasts, the peninsula of Devon and Cornwall and north-west Scotland, and crabs along the Yorkshire coasts.

Fish is regarded as one of the most suitable of our convenience foods; by that we mean that we can buy fish already processed so that it can be prepared and cooked easily and quickly. Fish fingers, fish cakes and fish steaks cooked with a prepared sauce in a plastic bag are examples of this. The fish processing factories which prepare these commodities are inevitably located close to the docks. *Why?*

A result of overfishing has been to encourage attempts to develop fish farming. There are a number of schemes now in operation particularly along the west coast of Scotland where sea lochs provide valuable shelter. Successful experiments whereby plaice and sole are farmed involve fertilising the eggs, hatching them in small tanks, transferring them first to bigger tanks and then to cages in the sea. In this way yields are assured and the fish are large enough to catch eighteen months after hatching. At sea they are not sufficiently mature to make a worthwhile catch until three years old. Fish tanks near the Hunterston power station produce particularly heavy yields because the water is warmer and this suggests possibilities elsewhere, as amateur fishermen on the Trent know.

**Rugeley Power Station with colliery.**
(*Photo: Patrick Bailey*)

**Section 4
ENERGY**

# INTRODUCTION

It is probably only when there is a 'power cut' at home and we are left helpless, wondering what to do and how to pass the time, that we appreciate how much we are dependent on electrical energy. Consider how your grandparents fifty years ago would have been affected by a 'power cut'. What equipment in your home is dependent on electrical energy? Certainly as our material standard of living has risen so has energy consumption.

Electrical energy is only one form of energy. In our homes electrical energy can be transformed into light, heat and sound: these are also forms of energy. The power stations which generate electrical energy may be thermal, in which case they use a fuel such as coal or mineral oil. Combustion of the fuel is a chemical reaction so that fuel is a source of chemical energy. Incidentally food is a source of chemical energy as well. Electrical energy is also generated by nuclear power stations and in this case the fission or splitting of an atomic nucleus produces heat and light. Hydro-electric power stations use the motion of water to turn turbines. Motion is a form of energy known as kinetic energy. Thus there are chemical, kinetic, electrical, light, heat, sound and nuclear forms of energy. There is also potential energy, a form of which, 'gravitational potential energy', will be considered in the hydro-electric section. We cannot make any form of energy; we can

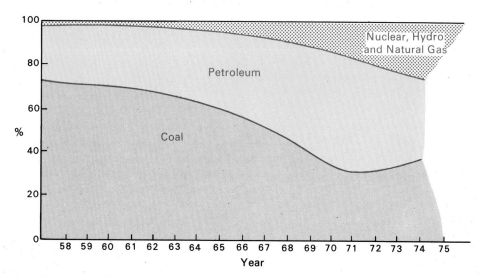

**Figure 4.1   The major contributors to the nation's supply of energy.**
The proportions are given as a percentage and calculated on the basis of 'coal equivalent', i.e. the relative calorific values of the other forms of energy when compared with that of coal.

only change it from one form to another, making it do something useful for us in the process. The key to all activity is solar energy; without it the world would be lifeless. From solar energy, for example, can be traced plant growth, including the swamp forests which were to be compressed to form coal, and also the weather cycle: evaporation, condensation and precipitation.

· The transfer of sources of energy from country to country contributes a large proportion of the total movements of commodities of the world. Fifty years ago coal dominated freight movements; today oil is the largest item by volume entering world trade. In the past, countries with readily exploitable energy resources had a great advantage. Much of the study of the 'Industrial Revolution' in Britain as elsewhere is about energy resources; as the early stages of the revolution progressed the essential requirement was coal and thus it is appropriate that we begin this section on specific sources of energy with a study of coal.

# THE COAL-MINING INDUSTRY

Towards the end of the eighteenth century there began in Great Britain what we call the 'Industrial Revolution'. The old systems of domestic industry, whereby goods were manufactured in the home, and handicraft industry, based on small workshops, were being superseded by the factory system.

Under the domestic system goods were produced by hand tools or simple appliances worked by hand or foot. Many of the first factories were located on the banks of fast flowing streams so that a water wheel could provide the motive power to drive a series of machines. The use of steam driven engines accelerated the emergence of the factory system.

Coal was not only vital in raising steam to drive machinery but in the improvement in iron production both as regards quantity and quality. The machinery being developed at the end of the eighteenth century was by present-day standards very inefficient. Early steam engines consumed between ten and twenty times as much coal per unit of energy produced as would be required by an efficient modern steam engine. Transport of coal to the traditional manufacturing areas of East Anglia, the south and the mid-west of England was slow and expensive. Consequently there was a massive relocation of industry and its concentration on the coalfields. A comparison of figure 4.2 and a political map of Britain will show that most of the large industrial towns are still those which are on or very near coalfields.

By the end of the nineteenth century the pull of the coalfields as a location factor was no longer so marked and yet new industry continued to be attracted, for there were the workers, the communication networks and the marketing organisations. It was not until the years after the Second World War that the pull of the coalfields for new industry was effectively countered on a large scale, although the trends were seen from 1930. In 1947, coal was still the major source of energy but it had become more convenient to use electricity generated in coal-fired power stations, and coal gas; consequently a much wider dispersal of industry was possible.

Since 1947 coal has faced intense competition from alternative sources of energy. In 1947 Britain produced 184 million tonnes of coal; now this figure has fallen to below 130 million tonnes and yet since 1947 consumption of energy has nearly doubled.

## The formation and types of coal

Coal is a hydro-carbon originating as rotting vegetation in the jungle swamps of Carboniferous times. The proportion of carbon, water, oxygen and hydrogen, as well as other materials, varies to give coal from different mining areas markedly different characterstics. The amount of pressure to which the coal seam has been subjected is a major factor in determining the character of the coal.

Most of the coals mined in Britain are classified under the broad grouping bituminous. A most distinctive type is steam coal, which has a high carbon content and on heating up breaks into small pieces so that combustion is rapid and great heat is given off: consequently this type of coal has been prized in the past because of its value in raising steam for ships and railway locomotives. No wonder decline in demand has been a contributory factor in the pit closures in the central parts of the South Wales coalfield, which is one of the areas noted for steam coal.

A particularly valuable coal is that which can be processed into a coke suitable for iron smelting. Such a coke should be light and as free as possible from impurities especially sulphur. It should also have strength sufficient to support the furnace charge. South Durham coal is an outstanding coking coal but most has been worked out.

Anthracite is a hard, brittle coal with a very high carbon content of the order of 92 per cent to 95 per cent. It is low in volatiles (oxygen and hydrogen) and burns slowly with little smoke and flame. This makes it particularly valuable in space heating and some aspects of food processing.

**EXERCISE**

To what extent and why has the demand for coals of various types changed during the course of the twentieth century? What reasons can you give to explain your answer?

## Mining methods

Mining can take various forms. Coal can be dug out from open pits after the overburden has been removed, a process called open-cast mining. When galleries are run into outcropping seams it is known as drift or adit mining and when a dipping tunnel runs down to shallow seams of coal it is known as slope or tunnel mining. The most usual, however, is shaft mining.

Mining profitability is influenced both by the physical characteristics of the field and also the previous pattern of mining employed. The dip and contortion of the coal seams and the occurrence of faulting can seriously handicap the extension of mechanised forms of mining. With thin seams it is necessary to remove 'country rock' to obtain the necessary height at the coal face. The condition of the rock overlying the coal seam, the roof rock, determines the character of the roof support which must be used, and the existence of gas or water can make further work difficult.

Traditionally coal mining has been associated with a great deal of human effort in a cramped, dust-laden environment. Mining operations developed into a three-shift system, one shift cutting the coal, one clearing it and one moving forward the roof supports and preparing the coal face.

Since 1945 much has been done in extending mechanisation in the mining industry. Mechanisation can be applied to three main

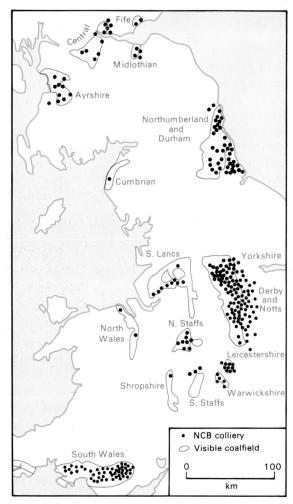

**Figure 4.2  Coalfields and collieries.**

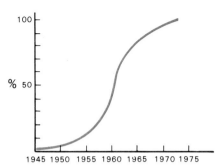

Figure 4.3  **Graph to show how the percentage of coal mined mechanically has increased over a period of thirty years.**

processes in the pit: the cutting of the coal, the loading and carrying of the coal to the shaft and the jacking up of the roof as the coal face is pushed forward.

There are various types of powered coal cutters and loaders but normally these machines cut between 1 and 2 metres into the face which extends for a distance of between 100 and 200 metres. The coal is cut and loaded on a conveyor belt in one operation. At the end of the run the cutter must be moved forward together with the conveyor belt and the powered jacks which support the roof. It is this sequence of tasks at the end of a run which provides so much demand on labour in a fully mechanised pit. When these

In what ways do the scenes which show pithead buildings and a coal face differ from those which might have been taken at a pithead and coal face at the beginning of this century?  (*Photos: NCB*)

processes are automated the saving on labour is considerable. However, the degree of mechanisation is dependent on the character of the seams. Seams which are level and of uniform thickness can be worked employing fully mechanised and automated techniques.

The great advantage of open-cast mining is the high production possible per unit of labour employed. This is largely due to the use of enormous diggers. With their use, a 20 per cent saving on the cost of extraction of open-cast coal compared to the average cost of underground mining can be achieved. Another feature of this form of coal mining is the greater adaptability of production to meet changing demands over a short space of time. The fluctuation in production is a result of a policy whereby unpredictable changes in demand are met by immediate adjustments in open-cast production.

The chief disadvantage of open-cast mining is the loss of land with its associated displacement of settlement and farmland together with the spoiling of the local environment. The Coal Board today ensures that land used for open-cast mining is restored to a productive state, usually as agricultural land.

## Organisation and trends

The National Coal Board (NCB) was established in 1947 and it then took control of 1400 mines. Many of these were long-established, small units working shallow

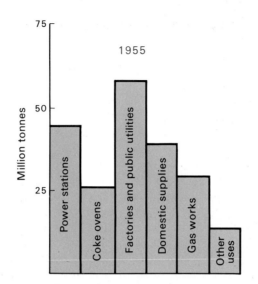
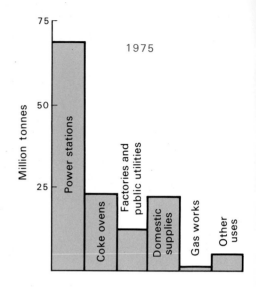

**Figure 4.4   The markets for coal 1955 and 1975.**

seams and these inevitably had such small reserves of coal that an expensive programme of mechanisation would have been hopelessly uneconomic. The mines that had reserves worth developing were retained, often expanded and deepened to take in new seams, and mechanised. Since 1947 a number of new, deep mines opened, for example in Nottinghamshire. This rationalisation of the industry whereby the mines with little future were closed reflects the competitive nature of the industry. By the early seventies there were under 300 working collieries.

Manpower in the collieries was three times as great twenty-five years ago as it is today.

Colliery closures and mechanisation in those retained are chiefly responsible for this. Ninety-three per cent of coal is now cut by power loaders. The reduction in manpower coupled with mechanisation has led to a marked increase in productivity per man-shift. In 1947 this was approximately 1 tonne of coal per man-shift: today this figure has more than doubled.

### EXERCISE

Make a list of markets that have decreased their demand for coal since 1955 and those that have increased their demand. For each, write a short paragraph to explain the reasons for the change.

86

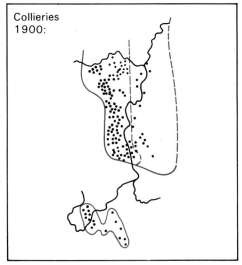

Collieries
1900:

Collieries
today:

Chesterfield

DERBYSHIRE

Derby

NOTTINGHAMSHIRE

Nottingham

LEICESTERSHIRE

Leicester

**Figure 4.5  Collieries in Nottinghamshire, Derbyshire and Leicestershire.**

## The East Midlands Division—A type study

There are two coalfields in the region: the larger is the North Derbyshire and Nottinghamshire, and this continues across the county boundaries into Yorkshire. The other coalfield is the Leicestershire and South Derbyshire. Although coal mining in this region can be traced back for at least 750 years, the great handicap to development for a long time was the difficulties facing the transport of such a bulky commodity. Most coal was carried by water; the shipment of Tyneside coal to London had been taking place for centuries. Since there was no navigable waterway close to most of the mining areas in the Midlands much of the coal was carried in bags slung across pack horses. This was very expensive: there are records dating back to the mid-eighteenth century which show that coal was costing up to ten times as much 30 kilometres away compared with the price charged at the pithead.

Expansion of coal mining in the East Midlands came at the end of the eighteenth century when canals were constructed to serve the area, but it was the coming of the railways in the period immediately after 1840 which heralded the great expansion of coalfields of this region.

Originally mining was concentrated in the western part of the Derbyshire and Nottinghamshire field where the seams of coal came near to the surface. The workings were shallow and numerous and much of the coal, at first, was dug from open pits. There still remain considerable quantities of coal in these western parts of Derbyshire but there has been a shift in production during the twentieth century to the eastern parts of the coalfield in Nottinghamshire. In this concealed part of the coalfield shafts are deeper and the collieries are much bigger units. We have seen how modern methods of mining, using elaborate machinery for the cutting and movement of coal, can be best applied in large collieries where seams are relatively fault-free, level and of uniform thickness. This is true of a great part of the two coalfields in this region, and accounts for the

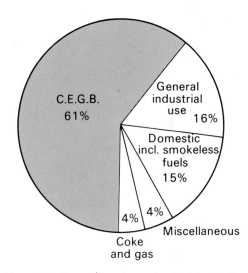

**Figure 4.6  The markets for coal mined in the East Midlands.**

87

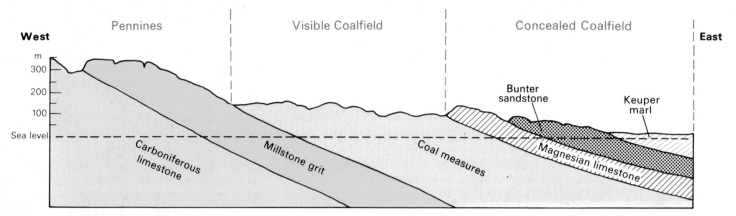

West — Pennines — Visible Coalfield — Concealed Coalfield — East

m
300
200
100
Sea level

Carboniferous limestone
Millstone grit
Coal measures
Bunter sandstone
Magnesian limestone
Keuper marl

**Figure 4.7    A section through the North Derbyshire and Nottinghamshire coalfield to the north of Mansfield.**

high level of productivity. Figures in excess of 3·5 tonnes of coal production per man-shift are now being realised over most of the region. Many new mechanical techniques have been pioneered here and include the long-wall retreat mining which we have already explained and the use of remote control equipment whereby a coal face can be worked by one man sitting at a panel of switches; this latter technique is still in very limited use. What problems face its wide-spread use in the mining industry do you think?

## EXERCISE

How does the pie graph (figure 4.6) differ from that which shows the national picture? Account for these differences.

# MINERAL OIL

Oil has a vast non-competitive market in much of the transport scene as well as in the rapidly expanding petrochemical industry. In the last quarter of a century it has also been able to compete successfully with coal as the fuel used for space heating, power stations, open hearth steel furnaces and various other industrial processes.

The great advantage that mineral oil has over coal is its ease of handling. It can be pumped, its flow readily controlled automatically: it can be conveniently stored; and there is no waste. In the mid-seventies mineral oil was the source for 46 per cent of Britain's energy demands.

Before the Second World War most petroleum products were imported in a refined state but since then conditions influencing the location of oil refineries have

changed. Since the Iranian oil industry, including the vast Abadan refinery, was nationalised in 1951 there has been a deliberate policy on the part of oil companies to establish new refineries away from producing areas that were economically under-developed and politically unstable. There are strong economic reasons for this policy as well. The refinery on an oilfield is tied to one source and with the enormous increase in production the productive life of a field is restricted. Before 1939 there was a great amount of waste in the refining process; even with very light crude oils only about half

What are the advantages and disadvantages in establishing an oil refinery at Fawley (a) to the Esso Oil Company and (b) to the community at large? (*Photo: Esso*)

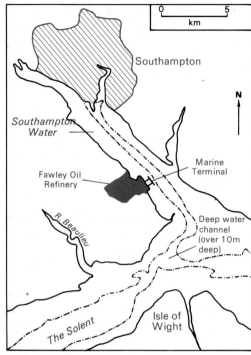

**Figure 4.8    Fawley Oil Refinery.**

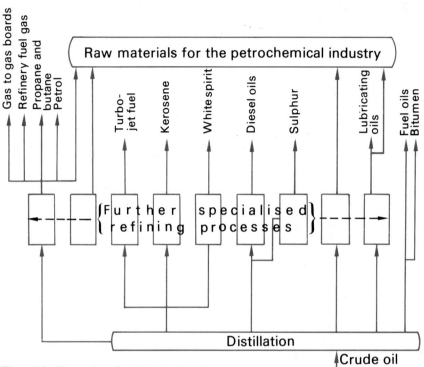

**Figure 4.9    The products from Fawley Oil Refinery.**

could be used. Today there is little wastage and at the same time much of the refinery production can be a source of raw materials for the petrochemical industry. Under these circumstances it is easier and more economical to transport crude oil rather than its refined parts. In the last thirty years there has been a fifty-fold increase in Britain's refining capacity.

## A study of Fawley oil refinery

Fawley oil refinery was established in 1921 but until 1949 it remained a very small concern. Between 1949 and 1951 the refinery was enormously expanded. It now employs 2200 people, a small number when one realises that the refinery covers 1300 hectares and has the capacity to refine 19 million

tonnes of oil in a year, but processes are highly automated. The main terminal, which has five deep-water berths and four coastal ones, can handle vessels of up to 300 000 tonnes: in fact more shipping tonnage is handled here in one month than by the Port of Southampton in a year.

Each day the refinery is likely to consume 60 million litres of fresh water, 2·3 million units of electricity and 3·5 million litres of furnace fuel (oil and gas).

There is a chemical complex associated with the refinery which produces feedstock for the manufacture of plastics, synthetic rubber, solvents, lubricating oil and fuel additives. Two local CEGB power stations use oil from the refinery and a Gas Board plant converts oil into synthetic natural gas (SNG) by a process known as reforming.

Fawley distributes its refined products by sea, using coastal tankers of 26 000 tonnes, and by pipeline (50 per cent and 41 per cent respectively) but smaller quantities are distributed by rail (7 per cent) and road (2 per cent).

**EXERCISE**

Draw the sketch map of the position of Fawley oil refinery (figure 4.8) on a large scale and annotate it as fully as possible using the information given above.

## Factors influencing the siting of an oil refinery

(1) The majority of refineries are located on the coast as 99 per cent of our crude oil supplies are imported or come from the North Sea.

(2) Transport costs can be cut by using super-tankers. Over 80 per cent of crude oil is transported in ships of over 50 000 tonnes. However, with ships of over 250 000 tonnes in operation and the tend for larger ships likely to continue, deep water facilities are essential.

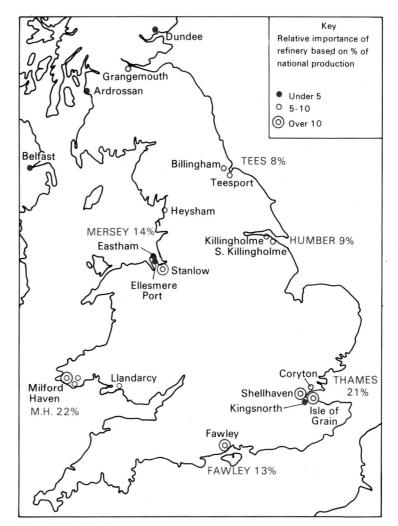

**Figure 4.10  Oil refineries.**

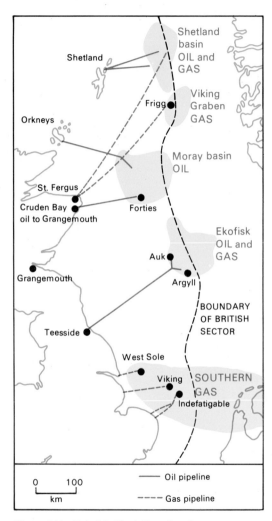

**Figure 4.11  Britain's North Sea oil and gas reserves.**

A sheltered berth in an inlet is also desirable.
(3) Most refineries are concentrated along estuaries near industrial areas.
(4) A large area of flattish land, perhaps 600 square kilometres, is required for the refinery installations. Reclaimed land is often used, as on Teesside.
(5) Refineries need to be isolated from the immediate vicinity of urban areas in case of accidents such as fire.
(6) The availability of a large labour force is not relevant in such a highly automated installation and the demands on fresh water and the supply of electrical energy do not restrict location at present.

## Britain's oil

In 1970 Britain's production of crude oil was just over 86 000 tonnes, which was equivalent to the annual output of a single Middle East oil well. The 1970s have been remarkable for the success of many exploratory drillings, particularly those off the Scottish coast and in areas adjacent to similar strikes in the Norwegian sector of the North Sea. As greater quantities of North Sea oil are brought ashore both by tanker and undersea pipeline, it seems likely that these oil fields will satisfy the domestic market of Britain. The impact of this indigenous source of energy is already being felt by the national economy and the effect on the oil industry has been enormous in some local areas such as the Shetlands and Aberdeen.

# NATURAL GAS

Today nearly all the gas consumed in Britain is obtained from gas traps under the North Sea and yet up to 1964 all gas used in this country was manufactured from coal and oil. In 1964 a contract to run for fifteen years was made between Britain and Algeria whereby natural gas in a liquefied form, at a very low temperature, was brought in to a special terminal at Canvey Island. From there it was piped to the most densely populated parts of the country. Before that time there had been speculation about the existence of gas fields under the North Sea.

In 1959 gas was discovered in Holland, trapped in a band of porous rock capped by deposits of salt, and the search for natural gas as well as oil was intensified in the North Sea. The Continental Shelf Act of 1964 gave Britain rights over about one-half of the North Sea area. Concessions were granted to several companies to prospect for natural gas on the understanding that the product would be marketed by the British Gas Corporation.

The first supplies of natural gas from under the North Sea were pumped ashore in 1967 from the West Sole field and these were distributed by way of the terminal at Easington in Yorkshire. Two more terminals were subsequently opened at Bacton, Norfolk, and Theddlethorpe, Lincolnshire, and a national transmission system,

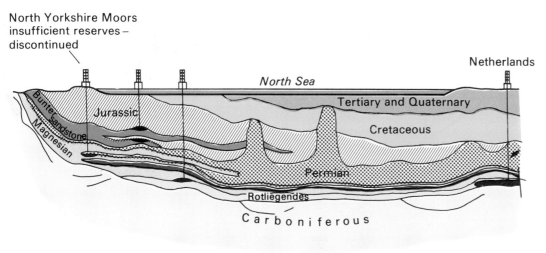

North Yorkshire Moors
insufficient reserves –
discontinued

Bunter sandstone

Magnesian

Jurassic

*North Sea*

Netherlands

Tertiary and Quaternary

Cretaceous

Permian

Rotliegendes

C a r b o n i f e r o u s

**Figure 4.12  Section across the southern North Sea to show natural gas traps.** (*Based on the Gas Council's 'British Natural Gas' theoretical section.*)

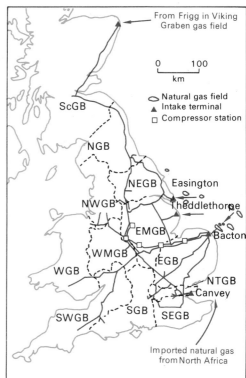

From Frigg in Viking Graben gas field

0     100
km

◯ Natural gas field
▲ Intake terminal
☐ Compressor station

ScGB

NGB

NEGB    Easington

NWGB    Theddlethorpe

EMGB

Bacton

WMGB    EGB

WGB

NTGB
Canvey

SGB   SEGB

SWGB

Imported natural gas from North Africa

**Figure 4.13  The natural gas grid.**

popularly known as the gas grid, has been rapidly developed. This feeds bulk supplies of natural gas at high pressure to regional distribution networks throughout the country.

## EXERCISE

Study the cross-section (figure 4.12) showing the idealised occurrence of natural gas deposits under the North Sea. Identify the geological conditions that have given rise to each trap, having studied the short account below.

In the same manner as coal and oil, natural gas was formed from organisms and plant matter which were covered by other layers of sediments and compacted. Natural gas can be extracted providing it lies in a geological trap where it has accumulated in a sufficiently large volume. The trap may be an anticline, a fault system or a salt dome. In each case the gas is prevented from escaping since it is in permeable rock with impermeable layers as base and capping.

# ELECTRICAL ENERGY

A power station, which is the popular name for a plant which generates electrical energy, does not store up and send out a load of electricity when required in the same way as a fuel such as coal, gas or oil. Electrical energy is generated in accordance with the demand at any particular moment during the day and a characteristic of the industry is the wide range in demand that it has to meet with in the course of twenty-four hours. Peak periods of demand are likely to occur between about 8 and 9.00 a.m. and 4.30 and 6.30 p.m.: periods of low demand occur between 11.00 p.m. and 6.00 a.m. Can you explain why this is so? Of course, seasonal demands vary as well. A frosty spell of weather will greatly increase the demand.

Figure 4.15 illustrates the enormous increase in the use of electrical energy during the twentieth century. This trend is associated with rising standards of living, reflected in domestic consumption, and the greater convenience of electrical energy with regard to both the domestic and the industrial market.

In the nineteenth century factories used electrical energy but many of them generated their own supply, particularly for lighting. Electrical energy is today available throughout the country by way of the national grid which is a unified system whereby electrical energy is transmitted at voltages of up to 400 000 over a nation-wide network. Control centres supervise generation to meet the constantly changing demand by shutting down stations during periods of low demand as required. In recent years there has been a marked relative decline in the use of coal.

Another trend is the decline in the number of power stations on the one hand and the enormous increase in generating capacity of new stations on the other: a modern power station can generate at a power of 1200 MW.

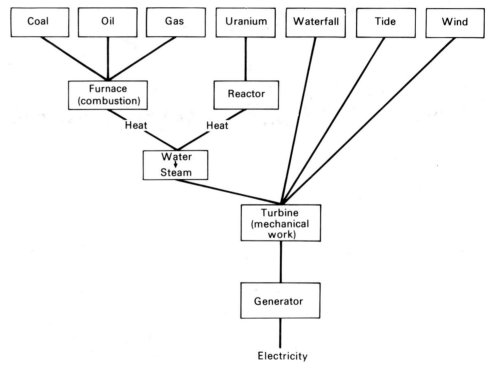

Figure 4.14  Sources of electricity.

## Thermal power stations

The general trends we have considered are reflected in the emergence of a concentration of new stations in certain favourable areas near to cheap supplies of coal and to large quantities of water. The cheapest coal is that from fields in Yorkshire, Derbyshire, Nottinghamshire and Leicestershire. Since these are inland coalfields the most appropriate source of water is the major rivers of the area, notably the River Trent but also the Aire and Calder. Oil fired power stations have tended to be concentrated in areas without coal, especially in southern England. Some have been sited close to oil refineries, as at Milford Haven and Fawley.

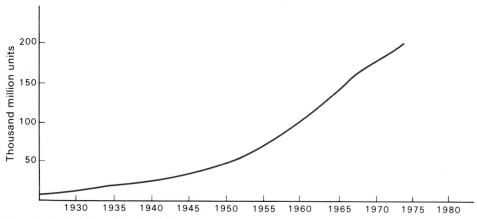

Figure 4.15 The demand for electricity.

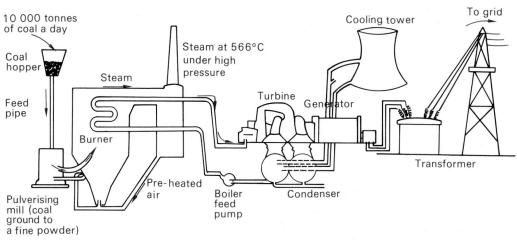

Figure 4.16 Processes in a thermal power station.

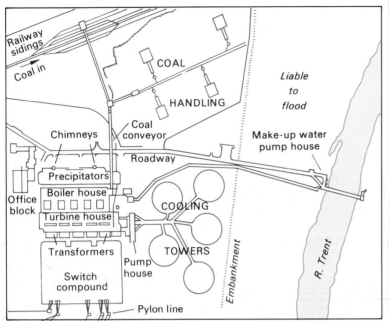

**Figure 4.17   Plan of High Marnham Power Station.**

▲ What location advantages has High Marnham Power Station? (*Photo: CEGB*)

## Study of the High Marnham power station

High Marnham was the first station to have a capacity of one million kilowatts. The station takes in 10 000 tonnes of coal a day when on full load. This is brought from nearby collieries in Nottinghamshire. The coal is broken down to a powder and blown, along with pre-heated air, into the furnace where it burns rather like a gas. The walls of the furnace are lined with pipes and the water in these is turned to steam. The steam is forced into the turbine motor which in turn drives the alternator which generates the electrical energy. The output steam is condensed in the condenser. It is here that so much water is used for cooling purposes. This coolant water then passes into the cooling tower. Some of this hot water is carried out from the water towers as water vapour and must be replenished by river water.

Study the following page and then identify the features which the two photographs illustrate. (*Photos: CEGB*)

Electrical energy produced in the generators is passed to transformers which 'step up' the voltage and the current can then be transmitted through the grid. The voltage will be stepped down again at substations before transmission to consumers.

# Hydro-electric power stations

The natural flow of water is globally an enormous source of energy but man's efforts to harness this energy have been achieved only at great initial capital outlay. Nevertheless, depending as it does on sunlight, it is a source of energy which will not run out—unlike coal and mineral oil. In Great Britain we have only a limited hydro-electric energy potential. The size of hydro-electric schemes is dependent on the volume of water together with the height of fall (known as the head). A major scheme can be based as much on a relatively small volume of water falling a long way as on a large volume falling over a barrage a few metres high.

Most large schemes involve the construction of a dam behind which a lake forms. Such a scheme has the advantage of creating a fall and providing a controlled flow of water from a reservoir. Diversion systems can be constructed whereby water is diverted from one catchment area into another to increase the volume of flow. Also diversion schemes can be used to lead off water from a natural water course to increase the head or fall of water.

Before a hydro-electric scheme is completed a number of very varied procedures need to be undertaken.

*Stage One.* A preliminary survey to identify a suitable site must take into consideration the annual rainfall as well as its seasonal distribution, the size of the catchment area, the amount of run-off (is the rock permeable or impermeable?) and the feasibility of dams, diversions and reservoirs.

**EXERCISE**

A deep, narrow valley formed in impermeable rock, free from fractures, is desirable. Why is this so? Give three reasons.

*Stage Two.* Once the scheme has been planned and before government consent is given, a public enquiry will probably take place and here objectors may state a case. One group of people who will be much involved are those who will be dispossessed of farmland, grazing or even homes, and compensation must be paid for any such upheavals. Hydro-electric schemes are located in the more remote, rugged and sparsely populated areas so that from an economic point of view compensation is not such a major factor in the costing of a scheme.

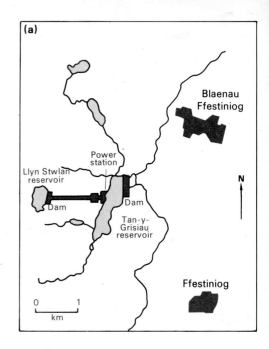

**Figure 4.18 Ffestiniog Power Station.** (a) location, (b) section.

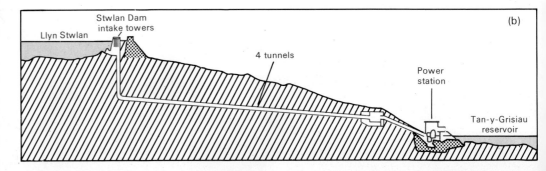

*Stage Three*. Remoteness suggests an inadequate communication system and before work gets underway on the scheme it is likely that roads will be built, or at least greatly improved, to cope with vehicles bringing in large quantities of steel and concrete. At the same time work camps must be built. Many more workers are involved in the construction of a scheme than in its operation.

*Stage Four*. Eventually there is the construction of the plant and the transmission system which will link it to the national grid.

Hydro-electric stations once established require only a small labour force to run and are not dependent upon expendable resources such as coal or mineral oil. A particularly valuable advantage over thermal power stations is the speed with which hydro stations can be brought on to full load; this means that they are used to provide a crucial backup to the thermal stations during periods of peak load.

## Type example—Ffestiniog, North Wales

The section, figure 4.18(b), illustrates the working of the Ffestiniog hydro-electric station. This is a pump-storage scheme. During periods of heavy demand on the national grid, it feeds into the grid but at off-peak periods water is pumped back into an upper reservoir. This scheme, which was built at a cost of £15 million, has a peak output of 320 000 kW. How does this compare with the Trent thermal stations? The need to preserve the amenities of Snowdonia National Park has been recognised by the planners and local stone has been used for buildings and extensive screen belts of trees have been planted.

# NUCLEAR ENERGY

The process known as nuclear fission was first recognised in the laboratory and published to the world early in 1939. An 'atomic pile' was first brought into operation in Chicago in 1942 as part of the programme for building an atomic bomb. It was not until after the war that serious attention could be given to the use of this new discovery for peaceful purposes. Then, in the 1950s, a shortage of coal together with a period of instability in the Middle East, which brought about some reduction in the imports of crude oil, gave a sufficient incentive to build nuclear power stations in this country. In recent years such stations have been seen as the long term solution to our energy demands. However, there have been recurring problems. The design of stations has been constantly modified. The fact that most of the fission products are highly radioactive has caused constructional problems concerning shielding and remote-control operations. It has also aroused public apprehension with regard to leakages and disposal of waste.

## The nuclear reactor

The fuel element consists of uranium bars about 2·5 cm in diameter and 1 metre long. These are sealed in containers of magnesium alloy. The fission of the nucleus is produced by neutron bombardment. When the nucleus is split it releases more neutrons, which in turn split other nuclei. Thus a chain reaction is set up and therefore conversion of nuclear energy into heat is maintained. The uranium fuel elements are embedded in graphite blocks which slow down the neutrons to speeds at which they are most effective. Without adequate control the reaction would very soon get out of hand: rods of neutron-absorbing material are therefore included in the design so that they may be inserted into or withdrawn from the reactor to keep the rate of fission at the desired level.

While the reactor is converting the energy of the nucleus to heat, carbon dioxide transfers the heat to a heat exchanger where water is converted into steam. The steam passes to the turbine and from there the processes in a nuclear power station are similar to those in a conventional thermal power station.

## Factors influencing the location of nuclear power stations (with special reference to Hinkley Point power station)

As with the other types of power station a great deal of planning is required before

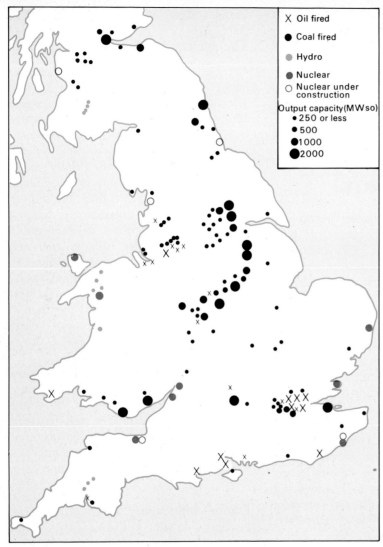

**Figure 4.19 Power Stations.** Account for the distribution of (a) oil fired, (b) coal fired and (c) nuclear power stations.

construction begins. Planners require considerable foresight to meet all the requirements satisfactorily and to ensure that the local amenities are not seriously impaired. Hinkley Point power station, which is situated on the Somerset coast on the south side of Bridgwater Bay, in fact consists of two stations.

Station 'A' has two magnox reactors which are capable of an output of 1250 MW. Nuclear power stations are likely to require twice as much water for condensing steam as comparable coal-fired plant. Hinkley Point 'A' alone uses 160 million litres an hour, a quantity so great that most inland sources of water would have been inadequate. Even a coast site can prove expensive to establish if there is a high tidal range since considerable construction work would be needed to ensure a regular inflow of water at all states of the tide.

Hinkley Point plant is on a 60 hectare site which is relatively level and free from flooding. Many coastal sites would offer these conditions but this site has a solid rock base which is able to support the enormous weight of the reactors which weigh over 60 000 tonnes each.

**A portion of the Petrochemical Division at Wilton.** (*Photo: ICI Ltd.*)

**Section 5**
**INDUSTRY**

# INTRODUCTION

We have so far been dealing with the primary sector of human activity which includes those activities directly concerned with exploiting natural resources; agriculture, forestry, fishing and mining are in this category. This section is concerned with the secondary sector, the conversion of raw materials into a manufactured form. Today two million people are employed in primary production and eight million in manufacturing. Compared with these figures it may be surprising to see that twelve million workers are employed in the tertiary or services sector, which includes transport, administration and commerce. However this trend, whereby an increasing proportion of people are employed in tertiary activities, is a characteristic of the twentieth century in most developed countries of the world. Nevertheless, we must appreciate that many people engaged in the service sector are actually employed by a manufacturing concern in assembling the raw materials, distributing the finished article and, not least, in selling it, so that these activities are each an integral part of the manufacturing scene.

# THE LOCATION OF MANUFACTURING INDUSTRY

There should be a best possible location for each individual manufacturing plant. One where production costs are as low as possible, is the 'optimum' location. The following section outlines the significance of locational factors.

## Raw materials

The influence of raw materials on the location of industry depends greatly on the weight/value ratio of the raw materials carried. A bulky commodity of low value per unit weight is likely to restrict the range of possible manufacturing sites. As the number of raw materials increases so does the difficulty in assessing their total influence as a location factor. At the same time the raw material of one industry can be the finished

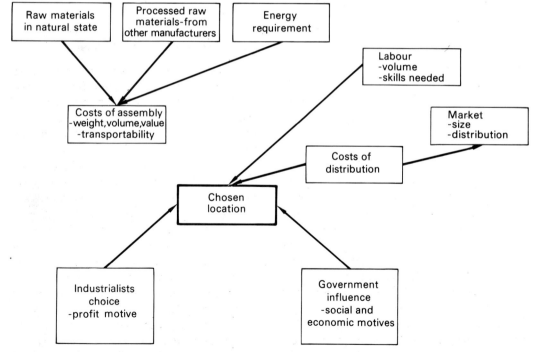

**Figure 5.1   Factors influencing the location of manufacturing plant.**

product of another. Perishability is yet another aspect to consider. It can have a profound effect, for example, on the location of certain types of food processing.

## Energy

Many long-established industries originally had only one form of energy available. Today there is likely to be a choice. The significance of the energy factor will depend upon the cost of the energy supply in relation to the total cost of the finished article.

## Labour

Availability and quality of labour are largely determined by the wages a company is prepared to offer. Such wages will be related to the skills of the workers and influenced by their bargaining power. For much of Britain's industry the value of 'traditional skills' has lost its significance since so many aspects of manufacturing today call for simple, repetitive manipulations which the worker can learn to perform. Many people today are still reluctant to move from areas where they have grown up, even if job opportunities there are limited. Nevertheless, in general more people nowadays do move away to work, and so the quality of the environment is becoming more important. The fact of an area being 'a nice place to live in' can make workers choose it in preference to other areas where work is also offered.

## Market

The marketing of the manufactured product can be a very complex process. The market may be a guaranteed one or it may be subject to enormous fluctuations in demand. For example the hosiery and knitwear industry is continually at the whim of changes in fashion and seasonal fluctuations in demand so that smaller firms in this particular industry can well be in danger of periodic redundancies and short-time working. On the other hand many food processing firms anticipate a guaranteed market for their produce from wholesale groups. Frequently there is reference to the term 'market orientated' industries. This trend for new industry to seek a site near to its market rather than near to the source of its raw materials is a major feature in industrial location in recent years.

## Human factors

In current studies on the location of particular firms it is being widely recognised that human factors (including the attitudes of industrialists, workers and indeed of the community into which the new industrial concern may come) play a vital part in location. At the same time, throughout the twentieth century there has been an increasing readiness for firms to shift from one activity to another. The beginning of the century saw the shift from carriages and bicycles to cars but since the Second World War there have

been many instances, for example from chemicals to textiles, and from textiles to plastics or engineering.

In the past, success often came because of the dynamic qualities of the factory owner. Today new developments are more likely to be planned with great care, involving detailed research into market, labour and transport conditions. There has been a growing demand in recent years for representatives of the shop-floor workers to be present at management level when detailed planning for the future is being discussed.

## The influence of government

In the past, much industrial plant has been established on a 'hit or miss' basis; success was often the result of a lucky choice: if the industry was sited in what turned out to be a favourable area it stood a chance of flourishing. An industrial area tends to attract new industry because of facilities there; another area might have been in a better position once the facilities had been created. It is now often considered to be a function of government to create these new facilities.

In Great Britain the intervention of government in the location of industry has become increasingly apparent since the thirties. The frequent changes in legislation on 'depressed' and 'development' areas since then shows a general concern for the languishing position of industry in certain parts of the country. The effect of government influence has been not to direct new industry to certain areas but rather to provide inducements. There have been the restrictions imposed on industrial building in congested areas, financial assistance to firms that consented to locate plant in areas vulnerable to unemployment and various forms of public investment in regions which are classed as depressed. The expansion of the vehicle industry in certain vulnerable areas of the United Kingdom is one example, but we will be referring to others.

## Site and local requirements

In recent years the industrialist has had to give much time and attention to the conditions offered by the site and locality when starting a new plant. This will involve not only a study of the area of building space required but estimates of room for expansion, storage space for raw materials and finished products and adequate parking space for employees' cars. The physical conditions of the projected site, including the drainage provisions, need for levelling, foundations and the availability of such facilities as gas, electricity, water supply, service roads and railway sidings must be considered. Some of these conditions will be more significant for certain types of industry. For example the demands on water will be very heavy when it is used as a coolant, an air conditioner or even as a raw material.

**EXERCISE**

Make a study of a nearby factory, considering the factors mentioned above.

# THE IRON AND STEEL INDUSTRY

## Background

The study of the growth of the iron and steel industry, together with the discoveries of new processes in making iron and steel, is most complex and not within the scope of this book. Nevertheless the significant steps should be known, for they help in understanding the broad features of the industry and explain some of the problems facing it today.

In 1709 Abraham Darby used coke made from Shropshire coal to smelt iron. There had been attempts before but Darby was more successful, probably because he used a bigger furnace and bigger bellows. There was an immediate expansion in this process but

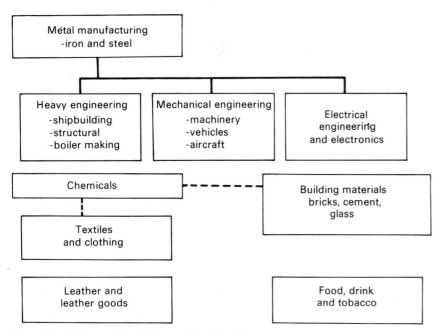

**Figure 5.2  Major branches of the manufacturing industry.**

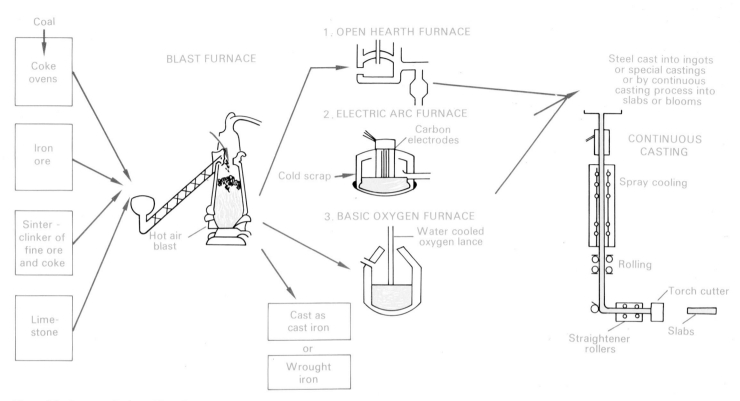

**Figure 5.3  Processes in the making of steel.**

still charcoal was to continue as a major fuel for smelting well into the nineteenth century.

In 1856 Henry Bessemer developed a process whereby air was blown through molten pig iron to burn off impurities. The alloy manganese was added to help deoxidise the metal. His process did not get rid of the phosphorus so that low phosphoric ores had to be used. However, in 1878 Thomas and Gilchrist showed that it was possible to remove phosphorus by adding limestone and lining the converter with firebricks containing magnesia and dolomite. Meanwhile in 1876 Siemens and the Martin brothers developed the open hearth process which soon became the most widely used. The process is much longer than the Bessemer (between seven and sixteen hours compared with twenty minutes) but larger amounts of scrap can be used and the product is purer. The distinctive feature of the process is that outgoing hot furnace gases can be re-used to pre-heat the incoming gaseous fuel.

## Locational factors and their changes

In the eighteenth century the most favoured location for an iron works was one near to iron ore found within the coal measures. These blackband and clayband ores were worked, for example, in the Lanark, South Wales and Yorkshire coalfields. At that time 8 to 10 tonnes of coal had to be converted to coke to produce 1 tonne of iron. Transport was so ineffective that a site away from a coalfield was inconceivable.

In the late nineteenth century as the blackband and clayband ores were exhausted, the Jurassic ores of Northamptonshire, Lincolnshire and South Humberside were being exploited. The relative leanness of the ore (generally under 30 per cent metal content) made it worthwhile to transport the coal to the ore, and so new works were established on the orefields.

In the twentieth century, production of domestic ores could not cope with the increased demand for steel and consequently imported ore, notably from Sweden, Spain, Canada and Australia, became more important. Thus new iron and steel works, as we shall see later, have tended to develop in coastal districts particularly where deep water terminals can be constructed to cope with bulk carriers.

## The manufacture of steel today

Figure 5.3 illustrates the various stages in the production of steel goods. The coking coal used should be as near as possible pure carbon with little sulphur and a low ash content. The resulting coke should be strong enough to support the charge of ore on top of it. Good coking coals are limited but that from the Bishop Auckland district of Durham proved of great value in the early stages of the Teesside steel industry although now it is largely worked out. South Wales and Yorkshire have good coking coals.

The character of a steel depends on the amount of carbon and alloys used in the steel converter. For high grade steels very carefully controlled conditions are needed. Electric furnaces, although costly to run, can provide these conditions. Another advantage of the electric furnace is that it can be loaded wholly with scrap, whereas the open-hearth charge is normally no more than half scrap, and only small quantities of scrap can be used in the Bessemer converter.

A major development from the Bessemer system is the L.D. basic oxygen process which was pioneered at the Linz and Donawitz works in Austria. This system is designed on a much bigger scale and can be controlled far more efficiently than the Bessemer, in fact it makes that and the open-hearth processes appear very clumsy to operate by comparison and, for this reason, the open-hearth furnace process is on its way out.

## Integrated steel works

A great deal of energy, mostly in the form of heat from fuel, is used in the production of steel, so the cost of this forms a major item in the final price of the product. An integrated steel works is planned to have all the processes on one site so that waste heat from one process may be used as input heat for a further process.

The refining of iron takes place at high temperatures and therefore a lot of heat is liable to be wasted. A crucial factor is the need to maintain the high temperature of the metal through its various stages of production. Pre-heating the blast in the blast furnace and also the gas intake in the converter is an economy. When the pig iron is run from the blast furnace it is kept hot in a huge reservoir known as a mixer; it can then be transferred to the steel converter in a molten condition. From there the steel ingots can be taken to the 'soaking pit' at white heat until they are ready to be received in the rolling mills. Here heat loss is rapid especially if the block is passed back and forth through the rollers, the traditional practice in British rolling mills. A continuous strip mill permits a continuous rolling through a whole series of rollers without a break, but such a mill is economically worthwhile only in a large steel plant with over one million tonnes production per annum.

The efficiency of the blast furnace and the steel converter is increased with size, partly because a lower percentage of the heat developed is lost and partly because the labour involved is reduced per unit of production.

Efficiency tends to increase with (a) an increase in the size of the plant and (b) the degree of integration of the production processes. In integrated plants it is possible to economise even further by providing coke ovens fired by gases from the blast furnaces and then using gases from the coke ovens to be burned in the converter, mixer and soaking pit, and also to generate electrical energy for rolling mills.

## Factors influencing the location of iron and steel plant

(1) *Site*. A modern steel plant requires a large site and for this reason new works tend to be located away from congested areas. A tidal site is valuable because of the greater reliance on imports of ore today, and the enormous water consumption necessary in steel production.

(2) *Energy*. The energy factor has already been referred to but note the wider choice that a steel works has; electrical energy can be applied in many processes and also natural gas used as a fuel. However, electric furnaces do use some coke as an element in the necessary chemical changes.

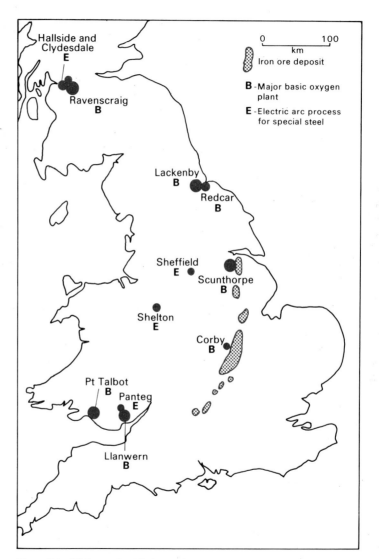

**Figure 5.4  Major steel-producing regions.**

(3) *Raw Material*. Besides the availability of ore, which was considered on page 106, scrap metal is a vital raw material. 'Home' scrap consists of ends and trimmings from the rolling mill and this might well take up one-third of the steel converter's charge. There is also 'bought' scrap which is normally obtained from dealers.

(4) *Transport*. Transport costs form an important part of the total delivered cost of the ore. In Britain transport costs normally account for about 40 per cent of the price of finished steel.

Specialist ore-carrying wagons have improved rail transport. Before 1939 sea transport relied largely on tramp steamers: today specialist ore-carriers of up to 150 000 tonnes play a leading role. Port improvements enable vessels to be 'turned round' more quickly. All such rapid bulk handling of materials is typical of recent developments in many fields and has reduced the overall cost of transporting and handling the ore.

## Organisation

In 1967 the British Steel Corporation was formed as a merger of the thirteen leading steel companies together with some 200 subsidiary companies. This new corporation has been concentrating on massive investment in installations of new plant at existing steel sites. Economies are being sought by increas-

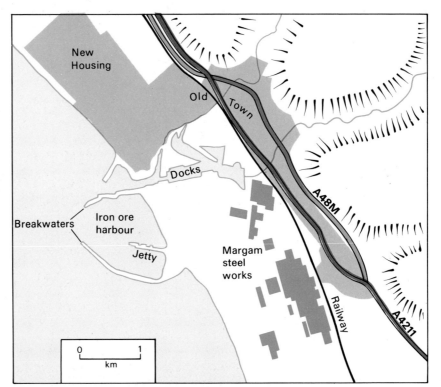

**Figure 5.5   Port Talbot.**

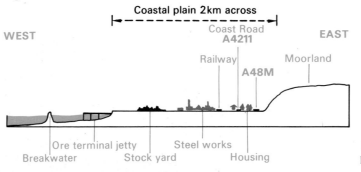

**Figure 5.6   Section across the coast plain at Port Talbot.**

ing the scale of production and closing down uncompetitive plant.

The problems facing the BSC were those which come from an industry with many small and scattered works, using outdated technical processes and old plant. Even when sited near the coast, port facilities were inadequate to cope with bulk carriers which meant that ore had to be transhipped at European ports to smaller vessels. As a result costs of assembly and production were high and productivity low.

The long-term aim is to concentrate production in a few areas where deep-water facilities are available to accommodate bulk ore carriers. Inevitably these changes create acute social problems. As older plant is scrapped either redeployment of labour or redundancies (or both) is inevitable.

## Port Talbot steelworks

The Abbey works at Port Talbot is one of the largest producers of steel in Britain with an annual capacity in excess of 3 million tonnes. A deep-water tidal harbour became operational in 1970 to cope with ore carriers of 100 000 tonnes. A conveyor system over 1·5 kilometres long carries the ore from the terminal to the stockyard. A little coal is im-

**Port Talbot Works.** (*Photo: British Steel Corporation*) ▶

ported but the majority is obtained from the South Wales coalfield. Two new basic oxygen converters now in operation are together capable of producing 3 million tonnes of steel a year. The operations of weighing, loading, control of oxygen lances, water-cooling pumps and steel tapping are controlled from a series of computerised instrument panels in a control pulpit in an adjoining building. Port Talbot is a principal supplier of thin sheets of steel to the motor car industry.

Similar developments are taking place at the Lackenby and Cleveland plant on Teesside, at Scunthorpe, at the Spencer Works at Newport, Gwent, and at Ravenscraig near Glasgow.

## Scunthorpe

Study the diagrammatic plan of the Scunthorpe region. The iron-ore deposits are the thickest seams of Jurassic ironstone in the country and the overburden is soft clays and shales which can be removed readily by huge mechanical diggers.

As extraction of the ore there has progressed the ore-bearing strata has become more deeply buried beneath overburden and also leaner in quality so that some mining rather than quarrying is becoming necessary. Scunthorpe deposits are today enriched by mixing with imported ores which have a higher metal content. Imported ore is brought to the recently constructed ore and

coal terminal at Immingham some 30 kilometres away.

After 1967 there was a general reorganisation of the industry in the Scunthorpe area. Besides the development of the Appleby–Frodingham works, projected before 1967, and the Redbourn and Normanby Park works, a new site on old ore workings which were infilled with blast furnace slag, was developed. This is the huge Anchor complex which includes ore-blending plant, sinter plant, continuous casting and basic oxygen steelmaking converters.

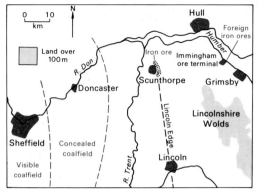

**Figure 5.7  The location of Scunthorpe.**

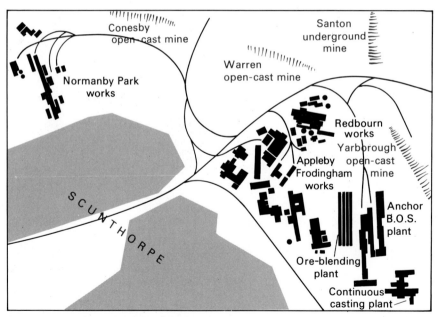

**Figure 5.8  The steelworks at Scunthorpe.**

# THE SHIPBUILDING INDUSTRY

## Locational influences

The Clyde was the major shipbuilding centre of the world and has launched more ships than any other inlet. That it achieved this distinction has been attributed largely to the expertise of the personnel employed there, workers and management alike; they had a start on their later competitors and built up not only a reservoir of skilled workers but also a reputation. However, the area needed certain vital conditions to allow the ship-building industry to flourish. The iron and steel industry which developed on the near-by Central coalfield provided the metal beams and plates and the Clyde itself is one of the most suitable, deep-water estuaries in Britain with ample space for expansion and sheltered water beyond for initial sea trials.

**Figure 5.9  Shipbuilding industry on the Clyde.**

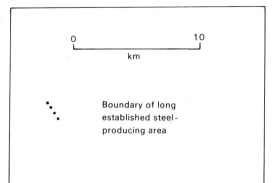

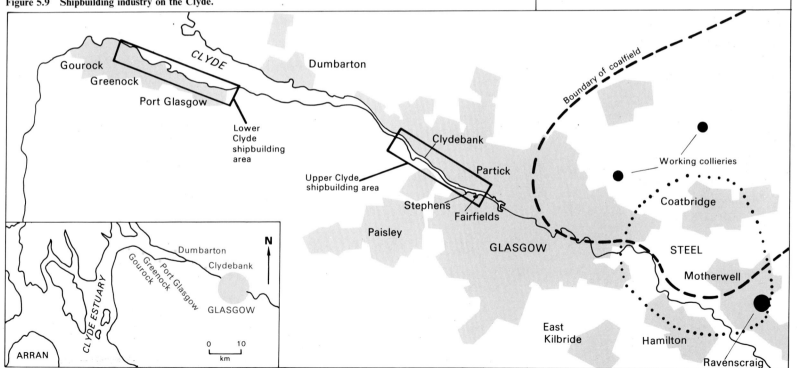

Britain's commercial interests, the colonies and its position as the leading mercantile nation of the nineteenth century created the demand.

In the paragraph above there are four broad considerations which must be recognised: the supply of raw materials, the site where the ships are to be built and launched, the people who build them, and finally the people who buy them. Today locational factors have inevitably changed in relative significance.

The trend, not only in oil-tanker construction but throughout the industry, has been for bigger ships. This places greater importance on deep waterway facilities being available adjacent to the shipyards.

Since many yards were first opened, the surrounding land has been developed for other purposes. The size of vessels built today was certainly not envisaged when the yard was first constructed and now the expansion of many sites is impossible. The increase in size of ship also means raw material supplies are very important; steel accounts for a greater proportion of the cost of a 100 000 tonne vessel as opposed to one of 20 000 tonnes. In the first case it could well be 50 per cent of the total cost and in the second 30 per cent.

Although it is impossible to introduce flowline techniques similar to those in the car industry, it has been those yards in the country which have built a series of similar type vessels which have made a substantial profit. In such cases production methods can be standardised and large subsections constructed, often by contractors outside the yard, in a continuous run over a period of several years. Today the reputation of the yard is as good as the last ship it delivered. A shipping company is looking for a fixed-contract job which will be quickly completed by a specific date.

For much of the twentieth century the industry has been handicapped by violently fluctuating demand; shipbuilding quickly reflects world economic trends and political crises. This tends to make it highly competitive, the successful yards being adaptable enough to maintain a full order book. Foreign competition, the decline of Britain as a colonial and trading nation, and the reduction in the number of ships being built, has forced the industry to rationalise.

On Clydeside, the shipyards are essential, both economically and socially, to the regions in which they are located and their collapse would cause severe problems in areas that already have above average unemployment rates.

To maintain Clydeside as a viable shipbuilding area, the government has been forced to take action frequently over the last two decades. Yards have been encouraged to amalgamate as part of a rationalisation programme which has led inevitably to some closures and a reduction in the work force. At the same time, huge sums of money have been granted to ventures which appear to have a chance of success in the face of foreign competition. A temporary easing of some of the industry's problems has occurred because of involvement with the oil industry and in particular with the construction of oil rigs.

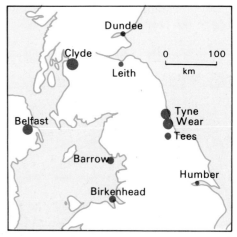

**Figure 5.10  Shipbuilding.**
Relate the factors which influence the location of shipbuilding areas to the distribution which is shown on the map. Consider the significance of an estuary site and the availability of sources of raw materials.

**Birkenhead. 1.** Describe the shipbuilding activities in the photograph. **2.** Explain how geographical factors have influenced the location of the industry. (*Photo: Aerofilms*)

# THE MOTOR CAR INDUSTRY

The industry began with little more than small workshops which were off-shoots from other manufacturing activities. The Humber, Riley and Rover firms grew out of bicycle and motor cycle manufacturing, Vauxhall was an engineering firm, whilst Wolseley was engaged in producing sheep shearing machinery. All these firms started with a very small output, as all the parts for their cars were made on the spot. Skilled labour was therefore essential, giving an advantage to the West Midlands area, which had a tradition in engineering.

In the early part of the twentieth century there were many firms in the industry, but only eleven produced more than a thousand cars each year. The introduction from 1908 of mass production techniques by Ford in the United States revolutionised the motor vehicle industry in Britain as well. Small scale producers were unable to compete with firms which achieved economies by large scale methods of production. As the optimum size of manufacturing plant increased, so the number of firms continued to decrease. Recent years have witnessed countless mergers, so that four companies now dominate the industry: British Leyland, Ford, General Motors and Chrysler.

## The nature of the motor car industry

(1) *Design*. The designer is influenced by a number of contrasting and sometimes conflicting factors. He is producing an expendable item which is likely to have a life of from four to ten years. He wants it to be reliable but only within this time span. He is in a field of keen competition where the outward appearance or styling of a car is as important as its performance: consequently he is influenced by the whims and vagaries of the buyer. Manufacturers know to their cost how some models, which failed to reflect the current mood of the market in design, were failures.

(2) *Tooling up*. The production of machine tools, which will make the components for a new model, is a most expensive part of the manufacturing process. A great saving is possible when a compromise is made whereby different models use many of the same basic parts. This is one of many advantages resulting from the amalgamation of companies.

(3) *Component factories*. The entirely different types of material which go into the production of a car require markedly different production techniques. Consider for example the production of tyres, springs, the pressed steel body, upholstery and sparking plugs. Therefore the development of specialist component factories producing large quantities of a limited number of car components was inevitable.

(4) *The assembly plant*. The assembly-line production method requires a high degree of organisation to ensure a steady flow of components coming in. The production processes no longer require the highly skilled mechanics needed at the turn of the century, but a large labour force of semi-skilled grades doing a limited number of repetitive jobs. Such a system, as we have seen repeatedly in Britain, is highly vulnerable to 'industrial action' from even small groups of operatives within the complex.

(5) *Marketing*. The methods employed in marketing are probably the most sophisticated used for the sale of any commodity, as might be expected of the most expensive single expendable item that most individuals are likely to purchase.

## The location of the motor car industry

Within the industrial states of the world today the motor car industry should have a very low location index: in other words it could be sited over a great range of locations. The influence of natural resources has been felt only indirectly.

Easy access to component factories by the assembly plant is vital. Assembly works carry very small reserves of parts, a 'hand to mouth' situation which soon shows itself when a component plant fails to deliver for a short time. Also since some component firms supply several manufacturers this will encourage a concentration of the industry into belts.

The massive success of a very limited number of vast firms again will lead inevitably to a great concentration of manufacturing in limited areas. The small scale firms have great marketing problems in competing with the established giants except in specialist lines—sports cars, high performance vehicles and three wheelers.

While these factors have tended to restrict dispersal of the industry and have underlined the attraction of a central location in the belt from London to Liverpool, the government has in recent years been able to encourage manufacturers to set up plant in areas of higher than average unemployment. Plants established on Merseyside and in Central Scotland received building grants and financial assistance. In fact plant influenced in such a way now accounts for 13 per cent of the total labour force of the British motor car industry.

## Motor vehicle manufacture at Dagenham

The plant at Dagenham belongs to the Ford Motor Company. Ford first manufactured vehicles in Britain at Trafford Park, Manchester, although initially the company was only concerned with assembling parts that had been sent over from the United States. As output increased and more parts came to be produced in Britain there came a

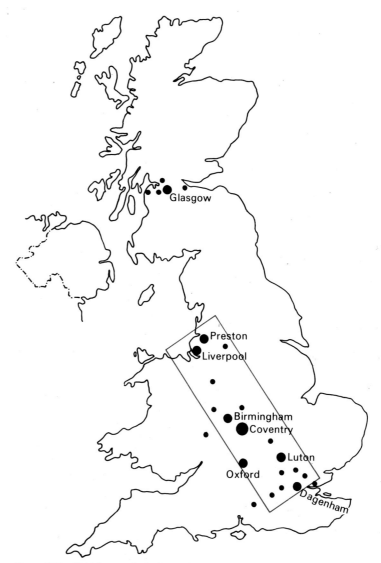

**Figure 5.11   Vehicle manufacturing centres.**

115

time when it was more economic to find a new site on which a purpose-built factory could be constructed, with room alongside for further expansion.

There were obvious benefits to be gained by moving south to Dagenham. A large market had developed in the South East of England, and this seemed likely to expand. Also by introducing mass-production techniques Ford had reduced their dependence on large quantities of skilled labour; and Dagenham, which was undertaking the re-housing of people from London's East End could provide plenty of unskilled workers. The Thameside site provided transport links by sea for incoming raw materials and outgoing vehicles. The 220 hectare site was mostly marshland, which had to be drained. However, it did offer room for future expansion. Some 2000 employees and their families moved from Manchester to Dagenham when the factory opened in 1931.

Iron ore, imported from abroad, and coal from the North East supply the blast furnace at Dagenham. The blast furnace, unique to South East England, produces high quality pig iron which is cast into cylinder heads, blocks and crankshafts in the foundry. Although there is a considerable amount of integration at the Dagenham plant the company still relies upon a large number of suppliers. Ford has been dealing with more than 5000 suppliers of raw materials and services. In an attempt to safeguard supplies as far as possible Ford has taken over some supply

plants such as Kelsey Hayes Wheel Company and Briggs Motor Bodies of Southampton.

Apart from acquiring smaller firms, the

company has established itself in other areas of Britain rather than expanding further at Dagenham. The first attempt at decentralisation, in line with government policy of the

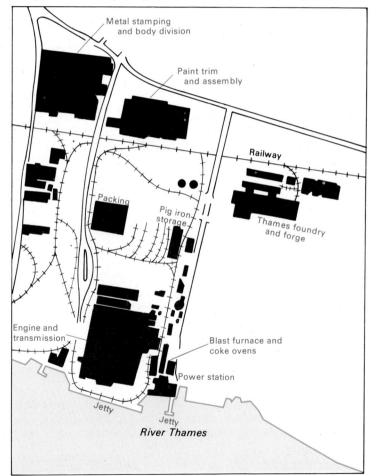

Figure 5.12   Fords, Dagenham.

**The Ford Works, Dagenham.** What locational advantages has this site? (*Photo: Ford*)

1960s, came with the opening of the Merseyside plant at Halewood in 1963. This was followed by a new tractor plant at Basildon, an axle and chassis works at Swansea, the Langley commercial vehicles plant, a service centre at Daventry and a research and engineering centre at Dunton. At the same time a new pattern of production developed. Decentralisation released valuable capacity at Dagenham and allowed the various Ford plants to specialise either in a limited number of models or in certain components.

# THE CHEMICAL INDUSTRY

## Introduction

The chemical industry is concerned with the conversion of raw materials through chemical reactions into large numbers of new compounds having a wide variety of different properties. Although the industry is a very complex one, much of its raw material consists of a very limited number of items, namely air, water, mineral oil, sulphur and common salt. There has been unprecedented expansion not only in total production but also in the variety and range of commodities produced. Although some of these products enter the retail trade directly (e.g. soaps, detergents, pharmaceuticals) most are used by other industries: it has been claimed that all industries rely to some extent on the products of the chemical industry. At the same time the distinction between a chemical industry and a chemical-using industry is often blurred.

An almost inevitable trend is the emergence of a few giant chemical combines. This reflects the interdependence of one branch of the industry with another and also the long and expensive research programme to produce new products in an intensely competitive field.

## Locational factors

*Raw materials* have influenced the location of the chemical industry, and for some sections of it their accessibility has been the dominant factor. The saltfields of Cheshire not only encouraged the industry to develop in the immediate vicinity, but were partly responsible for its growth in the Merseyside area. On Teesside local deposits of common salt and anhydrite and the local coalfield encouraged the growth of the industry while oil refineries such as that at Fawley, Hampshire, provided a source of raw materials for recently developed petrochemical works nearby.

*Communications* are also important. Since the chemical industry is dealing with bulky commodities, the advantages offered by tidal locations are frequently reflected in the siting of plant. The concentration of factories on Merseyside is an outstanding example. Study figure 5.13 and give three other examples.

However, the use of pipelines for transferring commodities does allow a greater degree of flexibility.

Since the chemical industry produces raw materials for most industrial activities its *market* is far reaching. However, there are many direct links between supplier and the industrial market. Bleach works and factories producing dyestuffs have been closely associated with the textile industries of Yorkshire and Lancashire.

There are other locational influences. Explosives factories are generally situated well away from populated areas, as at Penrhyndeudraeth in North Wales. As in the case of the car industry since 1945 government incentives have attracted firms to development areas where unemployment is higher than the national average. This encouragement helped establish the large chemical plant at Wrexham.

## Heavy chemical industry

The part of the industry which provides acids and alkalis for so many other manufacturing processes ranging from metallurgy to food-processing is the longest established.

Sulphuric acid is the most important industrial acid. It can be produced from sulphur and also minerals containing sulphur, such as iron pyrites and anhydrite, as well as being a by-product from coke ovens and gas plants. The acid is a bulky commodity, of

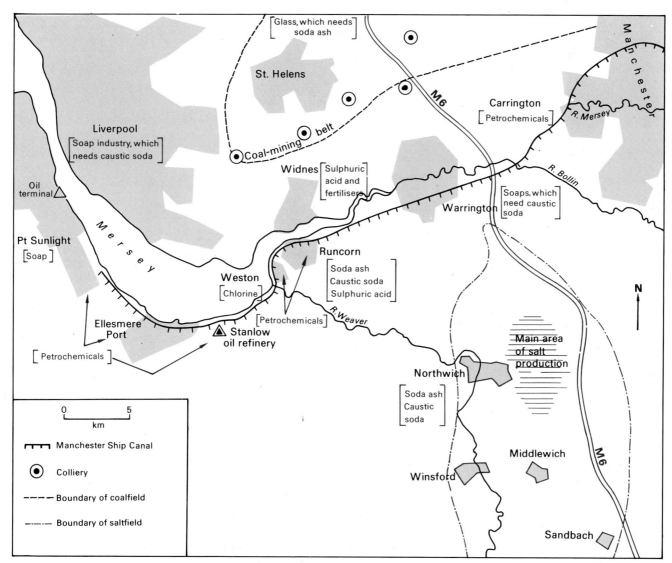

**Figure 5.13   The Merseyside chemical industry.**

low value and highly corrosive, so transport costs can make up a considerable proportion of the total market price. Sulphuric acid has many industrial uses not only in the chemical industry, where it is required in the production of explosives, plastics, dyes and carbonate of soda, but also in the iron and steel industry, in galvanising and tin plating.

Soda ash is the most important of the alkalis. Its production by the Solvay or ammonia–soda process uses common salt, normally supplied as a brine, together with limestone. Fuel requirements are heavy, so that their accessibility has often proved a dominant locational factor. Caustic soda, the other major alkali, is produced electrolytically from brine, and chlorine is a valuable by-product. These alkalis are vital elements in the production of glass, soap, rayon and paper. Materials for the manufacture of glass include fusing sand, lime and soda ash, while soap requires caustic soda as well as vegetable oil, fats, potash and chemical solvents.

## The petrochemical industry

The expansion of the petrochemical industry during and since the Second World War has revolutionised industrial activity generally. This expansion has been the result of concentrated scientific research which has led to the production of an enormous range of

◀ **Billingham.** (*Photo: ICI Ltd.*)

materials which can be used for example in the textile industry, building industry, and a wide range of consumer industries.

## Petrochemicals at Teesside

Teesside has been involved with the chemical industry since the Imperial Chemical Industries (ICI) established a plant at Billingham soon after the First World War. Billingham was chosen because it was on a sheltered tidal estuary and there was extensive, level, undeveloped land available. At the same time there were local deposits of common salt, gypsum, anhydrite and limestone besides coal from the nearby Durham coalfield. The anhydrite is mined from below the site of the works and is used, together with ammonia and imported potash and phosphates, in the manufacture of fertilisers. The production of ammonia is a major concern of the Billingham plant but other products include nitric and sulphuric acids.

Since the Second World War there has been a second stage of massive development by ICI, this time concerned with the manufacture of petrochemicals. A site was developed at Wilton on the other side of the estuary to Billingham. To supply raw materials for the petrochemical plant at Wilton, ICI combined with the Phillips Petroleum Company to develop their own refinery. To cope with this expansion as well as the attraction to the area of other petroleum and chemical concerns, the further reclamation

of mud flats in the estuary of the Tees has been carried out and the river itself has been widened and dredged so that it can accommodate oil tankers of up to 60 000 tonnes. Not only does the Tees provide facilities for shipping but it is also used as a source of water for cooling purposes in various chemical processes, and much noxious effluent is pumped into it.

There is a high degree of integration between the petrochemicals division of ICI at Wilton and other sections of the chemical industry. Ethylene, the main product of the Olefine plants that use naphtha, is consumed by the plastics division at Wilton. Some is even transferred by the Trans-Pennine pipeline to Runcorn where the Mond division converts it to vinyl chloride for use by the plastics section. At North Tees there is a plant that liquefies the ethylene in order to transport it in refrigerated ships to the Netherlands for the manufacture of polythene. Another product of the Olefine works is propylene, which is also consumed by the plastics division in the preparation of polypropylene, an essential raw material for the ICI fibres factory in Northern Ireland, where they make Ulstron fibre for blankets and

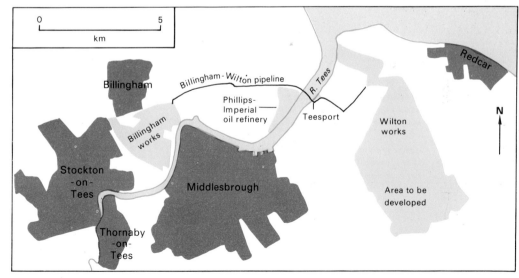

**Figure 5.14   Teesside and the chemical industry.**

ropes. Many more products emerge from this one division on Teesside, as intermediates to be used by other sectors of the company.

## EXERCISE

Jumpers made of acrylic fibres rather than wool, plastic cups replacing those made of paper—these are both instances of the increasing use made of petrochemical products. Provide more examples yourself.

**Figure 5.15 Major manufacturing plant which make up part of the Imperial Chemical Industries organisation.**

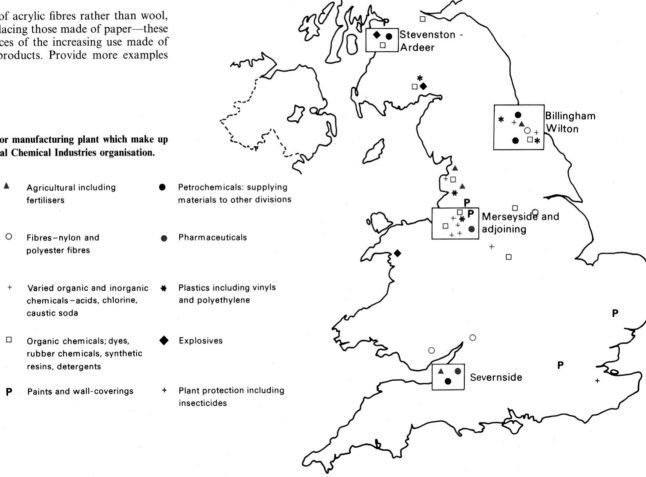

▲    Agricultural including fertilisers

●    Petrochemicals: supplying materials to other divisions

○    Fibres—nylon and polyester fibres

●    Pharmaceuticals

+    Varied organic and inorganic chemicals—acids, chlorine, caustic soda

✱    Plastics including vinyls and polyethylene

□    Organic chemicals; dyes, rubber chemicals, synthetic resins, detergents

◆    Explosives

P    Paints and wall-coverings

+    Plant protection including insecticides

# TEXTILE INDUSTRIES

Textiles are materials which are woven. They may be made from natural fibres such as wool or cotton, or from man-made fibres such as nylon or rayon. A third alternative is to use mixed man-made and natural fibres, often in an attempt to preserve some of the qualities of the natural fibre and at the same time reduce cost. Compare the price of an all-wool carpet with one which is made from mixed or completely synthetic fibres.

The textile industry has changed in character enormously over the last thirty years. This is largely due to the introduction of man-made fibres. Nevertheless, it is appropriate that we should begin this study with a consideration of the original textile manufacturing concentrations in South Lancashire and the Aire–Calder area of Yorkshire, not least because of the impact of this industry on the urban scene in these areas.

## Wool

A consideration of the factors which have influenced the location of the woollen industry gives a fascinating insight into the more general considerations in industrial location. The British woollen industry was well established in many parts of England by the fourteenth century. Although East Anglia was to emerge as the leading region, the distribution of the industry reflected very broadly the distribution of population. Thus although we can recognise the influence of

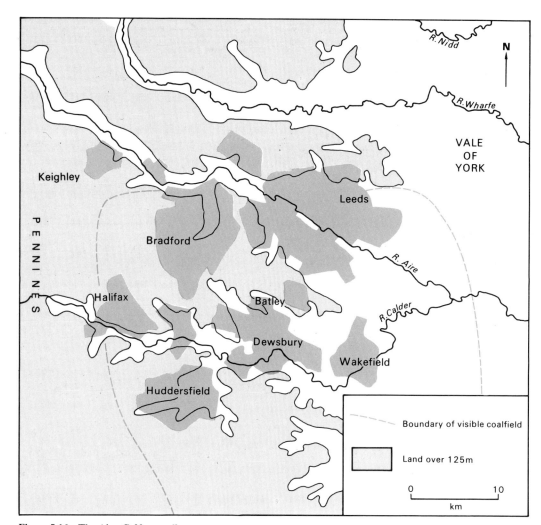

**Figure 5.16  The Aire–Calder woollen area.**

monasteries, of Flemish immigrants bringing in their skills, and of continental trade by way of ports of East Anglia and the south-east as factors in locating textile working locally, nevertheless the industry tended to be set up in the market towns of prosperous, well-populated agricultural areas.

The emergence of other locational 'pulls' came in the mid-eighteenth century. The use of water to power machinery gave an advantage to sites near to swift-flowing streams and thus there was some change in emphasis with greater expansion on the fringes of the Pennines. This was a short-lived situation however. With the development of steam-driven machinery from the last quarter of the eighteenth century, woollen manufacturing areas away from coalfields were at an enormous disadvantage because of transport costs for fuel and the difficulties of ensuring its regular supply at a time when cross-country transport was so inefficient. The final blow in many cases came when the industrial magnates of Yorkshire, already tasting success, were able to offer higher wages for skilled spinners and weavers than the smaller businesses in areas not so favourably endowed geographically.

### The character and quality of wool

Whereas the qualities of cotton, silk and many other fibres can be simulated by man-made fibres, wool has unique qualities the sum of which make it very difficult to replace

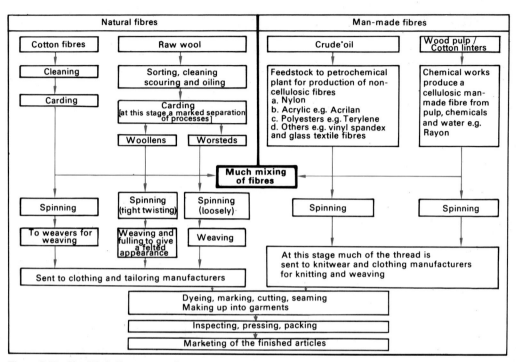

**Figure 5.17 Simplified diagram to show some of the processes in the knitwear and clothing industries.**

so that it has been more successful than cotton in competing with synthetics. Its resilience, elasticity, warmth and durability give it advantages in a wide variety of products ranging from carpets to suits. Nevertheless, new synthetic materials have recently been developed which more nearly provide the qualities which wool possesses.

### Cotton

The growth of the cotton industry in Lancashire had many features in common with woollen manufacturing but whereas the woollen industry has been able to maintain itself in general during the twentieth century for reasons we have already discussed, the

124

**Figure 5.18   The cotton area of Lancashire.**

Comment on the two scenes in a large knitwear factory. What do the photographs tell you about labour requirements? (*Photo: H. Flude and Co. (Hinckley) Ltd.*)

British cotton industry after reaching its zenith just before the First World War has had a subsequent history of periodic recessions and mill closures.

To enable us to understand these changes we should first consider the reasons for the growth of the Lancashire industry. There was a long-established woollen industry in the area, a plentiful supply of soft water which is valuable in all textile industries, the South Lancashire coalfield, the port of Liverpool, high humidity and the inventions of textile machinery in the area during the early stages of the Industrial Revolution which gave a particular advantage to this region. However, it is not sufficient to list a series of influences which could have been relevant in explaining the growth and ultimate supremacy of the Lancashire cotton industry. We must try to recognise the relative importance of these and view Lancashire in a world perspective.

The crucial point is that in the nineteenth century Lancashire had the start and Britain the commercial experience to enable the region to provide cotton goods for the world. With the possibilities for industrialisation on the South Lancashire coalfield, which the mill owners were quick to exploit, and the British Empire to provide the markets and contacts, such growth was possible. The British cotton industry was rapidly transformed from a local scattered series of mills using water power to the vast mills which steam power made possible. By 1913 Lancashire had built up the largest single exporting industry that had yet been experienced in the world—60 per cent of cotton goods entering world trade were from Lancashire.

The production of cotton cloth shows marked contrasts with that of wool. The enormous runs of uniform quality cloth are typical, which meant specialisation of spinning, weaving and finishing processes in separate mills, indeed in separate areas. This in turn allowed successful factories the much greater opportunity for continued expansion. The mass-produced article was able to capture the vast markets of the tropical world.

The apparent drawback of remote sources of raw materials has never been a dominant factor. Cotton fibre is non-perishable and its high value in relation to weight means that even when carried half-way round the world the transport charges amounted to only a small fraction of the total production costs.

Inevitably the bubble burst: competitors for our overseas markets were bound to come in. The 1914–1918 war (which upset our trading position) provided the opportunity, and after it Lancashire was never able to regain her position.

The recent trends have a familiar ring applicable to so many branches of industry. The cotton industry's competitive position has been maintained by scrapping surplus machinery and uneconomic units; less efficient firms have gone out of business and excess capacity has largely been eliminated. At the same time firms prepared to embark on major modernisation schemes have obtained grants from the government. Nevertheless, the industry continues to concentrate on the traditional area in Lancashire where over 90 per cent of British looms and spindles are located, with electricity now the source of energy.

The mills in Lancashire which have ceased cotton production have rarely been demolished. Some have been converted to producing synthetic fibres, sometimes retaining the old management but more often coming under the control of one of the big combines such as Courtaulds. Other mills have been adapted for a great variety of purposes including the manufacture of machinery, plastic kitchen ware, paper and a wide range of household products.

## Man-made fibres

It is sufficient to distinguish between, on the one hand, cellulose fibres which are produced from wood pulp and cotton waste and include rayon, and, on the other hand, nylons and other synthetic fibres produced by the petrochemical industry, the varieties of which run into several hundreds although many have characteristics which are very similar.

The first plants, not surprisingly, were established near existing textile industries, both cotton and wool. However, in recent years the industry has become very scattered as locational factors have proved less demanding. Energy from the national grid and low transport costs in relation to the price of the finished articles, together with the ease with which small units of the industry can be established, are the most significant influences. Thus there has been a dispersal of the industry for example to South Wales, Teesside, Northern Ireland and Grimsby.

While there must be the availability of raw materials, such as cellulose, and chemicals, such as sulphuric acid and caustic soda, any clear-cut cases in which a synthetic fibres plant has been established close to such supplies can be recognised only in limited areas: such an area is Teesside. However, the availability of a local supply of labour appears to be the major factor in many cases and government policy, which encourages industry to set up in areas liable to persistent unemployment, is very relevant here. There is no great advantage to be had from developing large production units capable of long runs of a single item, because the industry is subject to sudden changes in demand as new products are developed and become popular.

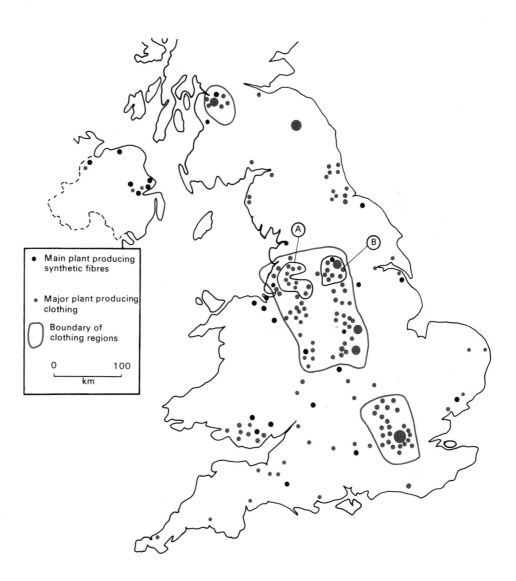

**Figure 5.19 Clothing and textiles.**
Identify the large centres concerned with clothing. What do 'A' and 'B' refer to? Describe the distribution of clothing plant and works producing synthetic textiles.

# FOOD PROCESSING

In Britain we are relying more and more on processed foodstuffs: that is, food which undergoes some manufacturing process between harvest and reaching the kitchen. As standards of living have risen so have more sophisticated forms of processing, and imports of food commodities have increased to supplement and provide alternatives to home produced foods.

We should distinguish between (a) unprocessed foodstuffs, both home produced and imported, (b) foodstuffs produced in and processed for the domestic market, (c) foodstuffs imported and then processed, and (d) imports of processed food. Give three examples for each group.

The food-processing industry provides some of the best examples of what are termed 'ubiquitous industries'; that is, industries which are very scattered, reflecting the distribution of the population. The advantages of a location near a market are clear-cut if there is an increase in the weight, bulk, fragility or perishability of the commodity as a result of the processing. Bread and confectionery are obvious examples. In these circumstances there is a tendency for the food-processing plants to be small since expansion would need a wider market and then distribution costs would become greater as well. The large bakery concern can overcome the problem by multiplying the number of its plants rather than expanding an individual factory.

On the other hand, perishability of the raw foodstuffs is likely to encourage the location of the processing plant close to the producing region. The location of pea canning and freezing plant in the Fens and Lincolnshire to process peas within an hour of picking is an example. Sugar-beet mills also tend to be within easy reach of the main growing areas but here transport costs are a major influence since the weight of the beet is much greater than that of the raw sugar produced.

The great concentrations of the food-processing industry occur at our major ports. Here imported foodstuffs can be more economically processed and packaged for distribution to markets throughout the country.

While some foodstuffs such as sugar, grains, beverages, butter and margarine must be processed to provide an edible commodity, others are processed with the aim of preserving them: the latter group include fish, fruit, vegetables and meat. All foods deteriorate with keeping but in recent years, with various sophisticated techniques, it is possible for many foodstuffs which would otherwise rapidly deteriorate to be preserved for long periods. This can be done by containering, freezing or dehydrating.

Processed foodstuffs are becoming more widely accepted as their advantages are recognised and as social conditions alter and standards of living rise. For example, concentrated orange juice is about one twenty-fifth as bulky as the fresh fruit needed to produce it; the consumer can be guaranteed standard of weight and quality with no waste; the enormous growth in demand for this commodity in developed industrial areas is a reflection partly of the demand for more balanced standards of nutrition and partly of the ease of preparation. We use the term 'convenience foods' meaning convenient to the consumer but here is a product which is also convenient to handle throughout the marketing processes.

Frozen products provide special problems, primarily to maintain the required temperature conditions not only when they are stored and transported but also when in the hands of the retailer and the consumer. The growth in the range and consumption of frozen products is thus a further reflection on a society where the refrigerator and freezer are accepted as necessities.

# MINERALS: THE BUILDING AND CONSTRUCTION INDUSTRY

In the past, buildings reflected the types of local materials available. Today old buildings particularly in rural areas frequently give an indication of local geology. We have seen, in an earlier section, how the light-grey Carboniferous limestone gives a distinctive appearance to the dry stone walling as well as buildings over large stretches of the Pennines. Similarly the mellow creamy brown Cotswold villages are particularly striking, but examples can be taken much further afield, particularly if you make a study of materials used in the construction of churches. Chalk is not a good building stone but flints, which are found in patches of the chalk, are often used in the construction of church walls in villages near downland countryside. The reddish tint of Triassic sandstone can be seen in many churches in Nottinghamshire and other parts of the Midlands. The Jurassic limestone has had the greatest reputation as a building stone. It has been sufficiently prized to justify its transport to many cities in south and east England.

The age of a building can often be recognised by the roofing materials. In the Snowdon region the towns of Bethesda, Llanberis and Blaenau Festiniog developed because of the presence of high quality roofing slate in that area and the slate-quarrying industry there became of great importance in the nineteenth century. Today evidence of this industry can be seen in the vast open quarries and spoil heaps which disfigure the landscape. With competition from

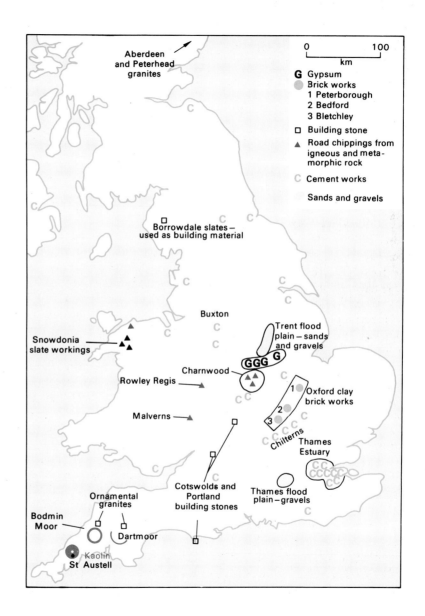

**Figure 5.20 Examples of major producing areas of materials for the building and construction industry.**

firstly clay and then cement roofing tiles the slate industry suffered a marked contraction and now quarrying is on a very small scale, the stone being used for ornamental purposes.

Twentieth-century buildings have generally lost a regional distinctiveness as brick and concrete have become the normal building materials. There are exceptions; can you think of any in your area? The production of cement takes place in many parts of the country but major producing areas reflect the influence of a bulky material and finished product whose transport costs form a considerable proportion of the price of the commodity. The leading area for the production of cement is on the Thames estuary. Give three reasons why you think it is located there (clues—water, market, rock). Other particularly important areas are the Chilterns, the Peak District, Billingham and Weardale.

Cement is made from a mixture of clay with powdered limestone or chalk; with the addition of water a paste is made and this is fired in a kiln and the hardened material is then ground up and a small quantity of gypsum is added. Gypsum was formed when layers of mineral salts were deposited on the floors of lakes as they dried up in Triassic times. Thus gypsum is quarried in various parts of the Midlands including the Trent Valley. It is also used in the manufacture of some types of plaster board.

The production of bricks has been concentrated particularly on a portion of the Oxford clay belt near the towns of Peterborough, Bedford and Bletchley. The clay here is thick and of an even composition. At the same time it has about 5 per cent carbonaceous material in it, so that fuel needed for firing is reduced. The clay is dug out from vast open pits after the overburden of superficial deposits and too-loosely consolidated clay has been removed. The clay is then ground, pressurised into the brick shape in moulds and then fired in a kiln.

Apart from washing and grading, very little preparation is required for gravels and sands which form a vital raw material for use in concrete and for the foundations of roads. Deposits are widespread and therefore extraction tends to be near where there is a demand. Some of the most extensive workings are located on the gravel terraces west of London. Here open pits mark areas where gravels have been removed leaving a floor of the underlying impervious London clay. Inevitably this has filled with water and provides sites which can be used for infilling with rubbish or developed as facilities for fishing and boating. It is unfortunate that so many of the 'wet pits' still remain as useless, ugly marks occupying what was once some of the country's most fertile agricultural land.

Tough igneous and metamorphic outcrops such as granite are ideal as road chippings. Again the distribution of workings reflects not the size of the outcrop but the demand. Thus small outcrops in the Midlands including several workings in Charnwood and the neighbouring area have a far greater production than the more isolated igneous masses of Devon, Cornwall, Wales and Scotland.

## Kaolin

It is appropriate at this stage to consider Kaolin or China Clay. This is produced as a result of the decomposition of felspar in granite. There are large accumulations of this decomposed rock around the edges of a number of granite masses in Devon and Cornwall especially in the St. Austell area. The Kaolin is worked in deep, open pits by 'hydraulic sluicing'. The pits are surrounded by great dumps of waste matter. Kaolin is used in the manufacture of quality china, in the textile industry and in the production of certain types of high quality paper.

# SERVICE ACTIVITIES

We can divide human activity into three broad groups. Primary activity is concerned with such things as farming, lumbering, mining and fishing. Why should these be classed together? Manufacturing is a secondary activity, the conversion of raw materials into processed commodities. The tertiary group of activities is given the broad term 'services'.

Within this sector are very varied occupations which can be classified as follows.

(1) Financial including banking and insurance.

(2) Professional and scientific jobs which include those involved with education, law and medicine.

(3) Administration and defence, a group including the civil service, local government and the armed services.

(4) Public-utility services include the provision of gas, water and electricity.

(5) A group of activities classed as cleansing and protection; this includes the fire and police services as well as those concerned with waste disposal and sewage.

(6) The distributive, marketing and retail trades.

(7) Catering and entertainment.

Figure 5.21 shows the total number of persons engaged in service industries as a proportion of the total number employed. You may be surprised at the size of this group compared with the totals in manufacturing and primary activities. The size reflects the present sophisticated standard of living. In many of the underdeveloped nations, for example in most central African countries, tertiary activity involves less than 5 per cent of the estimated working population.

The distribution of people employed in service industries should to a considerable extent reflect the general distribution of population but some of the categories will be concentrated on certain focal points. The over-concentration of both government and commercial offices on central London has created drawbacks; competition for office space makes this a high-cost area. As more and more people came to work in London and, of necessity, had to find housing in the suburbs, traffic congestion during the rush hour became a major problem. In recent years these difficulties have led firms and government departments to find new premises away from central London. To a lesser extent similar problems are often encountered in the larger regional centres.

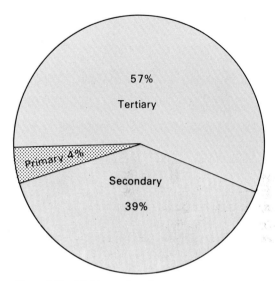

**Figure 5.21  Divisions of employment and the tertiary sector.**

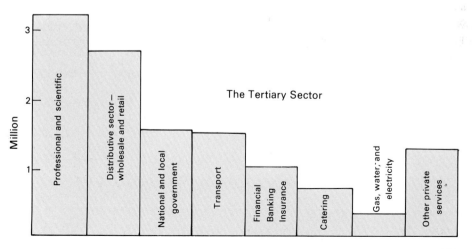

**EXERCISES**

(1) Consider the seven groups in the classification of tertiary activities. Write a paragraph on each explaining the likely distribution of people employed in that category. There may be a greater concentration of some activities: (*a*) in the capital city, (*b*) in the county town or major regional centre, (*c*) in the local market town, (*d*) evenly distributed throughout all well-populated areas, (*e*) having a more specialist distribution.

(2) Between 15 and 20 per cent of the country's working population are employed in distributive and marketing trades. Retailing is the provision of products to individual members of the public. This may be the end of a chain of transactions as products are transferred from the primary producer to the manufacturer to the wholesaler and then to the retailer. What do you understand by the following terms: (*a*) the 'shop round the corner', (*b*) the 'High Street' shop, (*c*) the shopping precinct, (*d*) the street market, (*e*) the suburban hyper- or supermarket? What advantages and disadvantages do each have? What trends can you see in shopping habits in your area over the last five years?

◀ What are the advantages and disadvantages of these two distinct forms of retailing (a) for the shopper, and (b) for the retailer?

# WATER SUPPLY

A reliable supply of fresh water has been a major influence on the location of settlement in the past. Before piped water was available, the site of a village often had to be beside a shallow well or spring and a village was likely to consume 500 to 1000 litres of water a day. In recent years, towns the size of

Southampton, Cardiff, Nottingham and Newcastle are each consuming over 100 million litres per day. Not only is consumption on a much vaster scale, but standards of hygiene are much higher today, although pollution of many of our sources of water supply is far greater, so that an elaborate filtration system is necessary.

Projected figures of consumption suggest that demands will continue to increase so that consumption could well be more than double the present figure within the next twenty-five years. At present each person uses, on average, over 150 litres of water per day. As standards of living rise, so consumption tends to increase: people take a bath or shower more frequently; more buy washing machines and dish washers; hosing down the car and using the sprinkler on the lawn are now commonplace; you can probably think of other examples which result in a further increase in the use of water.

We are liable to forget that demands from industry are increasing also. Power stations have particularly heavy demands and although water can to a certain extent be re-used there is a loss of 40 to 50 per cent in each cycle (refer to the section on thermal power stations). Chemical works, the iron and steel industry and food-processing concerns also rely on large supplies of water. Most manufacturing towns consume at least as much water for industrial purposes as for domestic.

More water is used in farming today. Dur-

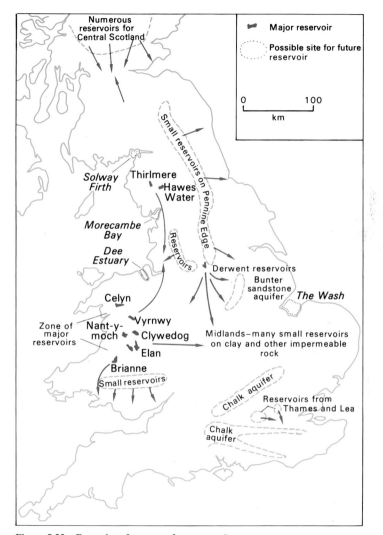

**Figure 5.22  Examples of sources of water supply.**

ing most summers especially in the south-east of the country it proves worthwhile to use spraylines to water crops such as potatoes and vegetables.

One of the biggest problems in planning future supply schemes is the unreliability of rainfall. Although our water supply can be obtained in various ways, ultimately it is dependent on rainfall and, as our resources are more and more stretched, spells of dry weather cause increasing concern.

## The sources of our water supply

(1) The most reliable source of water at the present time is the reservoir located in an area where rainfall is heavy, the catchment area is extensive, the rock relatively im-permeable so that surface run-off is consider-able, and the relief such that a potential site for a dam and reservoir can be found. The most accessible sites which provide these requirements are in Wales. The Elan reser-voirs on the upper Wye supply 200 million litres of water per day to Birmingham. The impermeable grit areas of the Pennines have smaller catchment units. Here the Derwent reservoirs provide some of the water supplied to Sheffield, Derby, Nottingham and Leicester.

Natural lakes can be used as reservoirs particularly if the water is raised by some artificial embankment at the outlet. Hawes Water and Thirlmere, which provide water for Manchester, are examples. In Scotland there is no immediate problem on the scale that English water boards are facing. There are a number of reservoirs in the southern Grampians as well as the Southern Uplands which supply water to the central lowland belt and there are new schemes being developed which will cope with the projected increase in demand.

(2) River water has for long been used as a source of domestic, industrial and agricul-tural supply. Irregular flow has always been a restriction and added to this, during the present century, there is pollution. Never-theless the River Thames is still the major source of water for London. In this case water is channelled from the Thames to vast reservoirs dug out of the gravel to expose the underlying impermeable London clay (refer to the map of west London, figure 5.24).

(3) A third source is underground water ob-tained by sinking wells or boreholes. Some rock is very permeable and will hold vast quantities of water either in pores or in fissures; we know this rock as an aquifer (water carrier). Chalk is one such rock and Bunter sandstone another. The chalk under-lying the London Basin has been tapped to such an extent that over-use has lowered the level of saturation considerably. Bunter sandstone is a valuable aquifer for example in parts of the East Midlands particularly Nottinghamshire. In the future it may be possible to regard aquifers as controllable reservoirs, 'topping them up' when necessary.

A more obvious site for reservoirs is the great coastal inlets which, if enclosed, could become vast fresh-water lakes. The Wash, Solway Firth, the Dee estuary and Morecambe Bay have been studied with this in mind although as yet expense has outweighed the incentive of demand. Desalination schemes using sea water are possible ways of increasing supply but would be even more expensive than the coastal inlet projects for the scale of schemes needed to cope with the likely increases in demand.

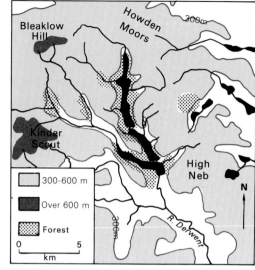

**Figure 5.23  The Derwent reservoirs.**

**Figure 5.24  Wet pits and reservoirs west of London.**
Explain the significance of the features on the map. Built-up areas are marked in a red tint. Some filling in of wet pits is taking place for example in the Ashford area.

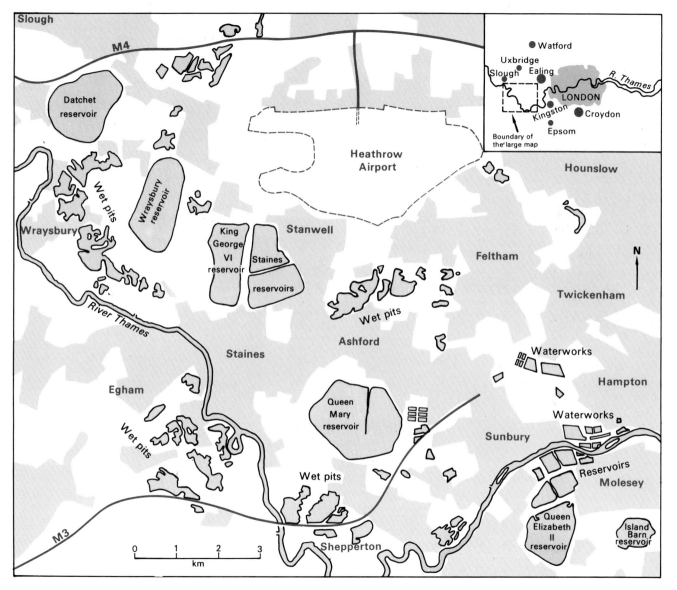

**Brickworks at Ridgmont, Bedfordshire.** Suggest three locational advantages that this brickworks has? (*Photo: Aerofilms*)

**Dinas Dam, Cardiganshire.** What is the purpose of this dam and why do you think it has been built here? (*Photo: Aerofilms*)

**Gravelly Hill Interchange (Spaghetti Junction).** (*Photo: Aerofilms*)

Section 6
TRANSPORT

# INTRODUCTION

The efficiency of the nation's transport system is clearly linked with its economic prosperity. It must cope with the regular daily commuter traffic and also the less well-defined movement of business, commercial and sales representatives, as well as passengers travelling for social purposes. The vital feature of passenger transport is the need to deal with the violent fluctuations in use on both a daily and a seasonal basis.

As for the demands of freight transport, the cost of the assembly of the raw materials and the distribution of the finished product to the market can be a large proportion of the total cost of the commodity. The choice of mode of transport will depend on cost which is based partly on a 'weight to distance' ratio but also frequently on complex privately negotiated contracts. Speed may be crucial, particularly with perishable items, while the available systems for handling, storing and packaging have become major factors when the customer makes his choice of which form of transport to use.

Much of the costing involved in transport rates is associated with loading and unloading. Any 'break-in-bulk' or transfer from one form of transport to another is likely to bring a considerable increase in the costing. A commodity that can be poured or pumped has obvious advantages.

The provision of uniform sized boxes or containers can accelerate loading and pack-

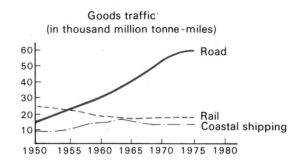

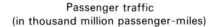

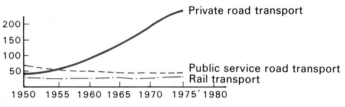

**Figure 6.1  Freight and passenger transport.**

ing and allow new techniques including automatic processes to be developed so that packages are readily interchangeable between road, rail, sea and even air; the result is a faster, cheaper turn-round of goods in transit.

# ROAD TRANSPORT

Road patterns of the present day are frequently based on those established cen-

turies ago. This legacy of the past may be tolerable in a local rural or suburban setting but it is not conducive to efficient movement by through traffic today; and yet only in the last twenty-five years has the arterial road system of Britain begun to break away effectively from the medieval pattern.

This break has been achieved firstly by improving the courses of existing roads and secondly by the construction of new highway systems.

## EXERCISE

Study the maps (figures 6.2 and 6.3) and the photograph on page 137, and then answer the following questions.

(1) What is the function of a motorway as opposed to other roads and what are its advantages and disadvantages?

(2) How do the detailed plans (figure 6.3) and photograph on page 137 illustrate this function of the motorway?

(3) What are the problems facing both the planners and engineers before a motorway such as that illustrated in the photograph can be constructed?

The great advantage of road transport is its flexibility. The driver may travel when and where he pleases and at his own speed, within limits! However, the expansion of road transport is not due only to flexibility. It is partly a response to a rise in living standards and material well-being whereby the private car is no longer considered a luxury but an essential by most families in the country. At the same time the amount and proportion of freight which is carried by road transport has mounted at an extraordinary rate in the last twenty-five years. Today over 80 per cent of all inland freight is carried by road.

The improvement in design and increase in the numbers of vehicles together with the increase in carrying capacity of freight vehicles have simultaneously encouraged improvements to the road pattern and been assisted by them. Paradoxically flexibility of road transport has led to the emergence of many problems. Road traffic saturation (a traffic jam) is largely the result of a recurring

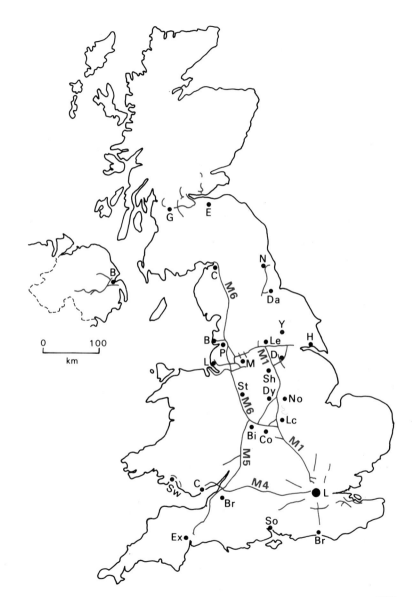

**Figure 6.2   Motorways.**

dilemma in transport planning, namely the uneven demand on roads which have not been designed to cope with the volume of traffic using them at peak periods. Road traffic cannot normally be rationed on any stretch of the highway in the same way that trains are on railways. Commuters daily travel to and from work creating the rush-hour situation while the seasonal pressure on roads leading to popular holiday resorts can lead to many hours of frustration.

In congested urban areas the private car is inevitably an extremely wasteful means of transport, compared with bus or suburban train. The street pattern in many cities seems no more able to cope with motorised traffic than the meandering country road but the expense involved in making road im-

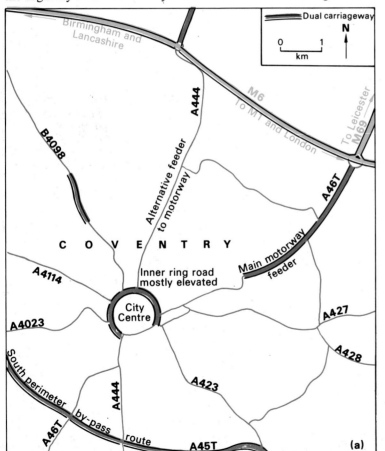

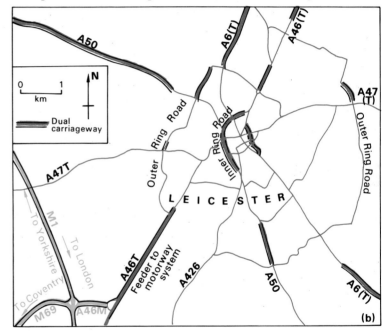

**Figure 6.3 The major road patterns for (a) Coventry and (b) Leicester.** Compare the road patterns of these two midland towns.

**Container Terminal, Harwich.** Describe the scene. In what ways has the container system revolutionised the transport industry? (*Photo: Patrick Bailey*)

provements is enormous. Nevertheless, elaborate road systems are being developed to permit free flow in many of the major cities. At the same time multi-storey car parks have been constructed to solve the parking problems.

The alternative to improving facilities for traffic near to the city centre is to impose various forms of restriction. This can be done by developing traffic-free precincts, by limiting parking, or by creating restrictions on admittance (for example permitting access to public transport and goods vehicles only).

# RAIL TRANSPORT

In Britain the most active railway building period was between 1840 and 1870. It was a period of intense competition between the railways and canals as they vied with each other to serve the new industrial areas. Railway companies frequently constructed tracks alongside canals and even bought up canals in order to let them fall into decay. The companies themselves competed with each other so that many tracks were constructed which duplicate the services for an area.

Throughout the second half of the nineteenth century railways dominated other forms of land transport and it is only since the Second World War that their position has been challenged by roads. It was in the

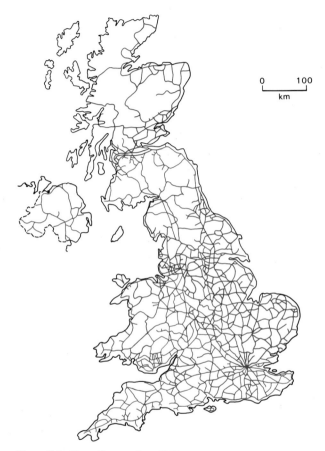

**Figure 6.4 The railway pattern 1950.**

**Figure 6.5  The railway pattern 1978.**

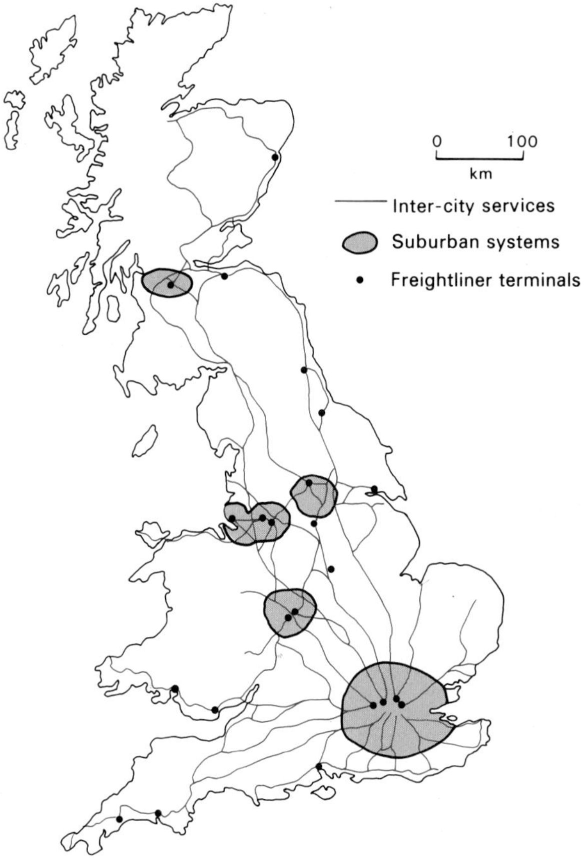

**Figure 6.6  The three main concerns of British Rail.** (a) The inter-city services. (b) Freightliner services. (c) Suburban services.

Inter-city services

Suburban systems

Freightliner terminals

fifties that the railway system became increasingly uneconomic as the cost of maintaining many hundreds of miles of little-used track and old signalling systems mounted. By the end of the 1950s British railways were facing an annual deficit of over £150 million a year. Inevitably in the last twenty years there has been the elimination of duplicate lines as well as many miles of rural routes which were proving hopelessly uneconomic to maintain. At the same time, from 1955, a modernisation plan was set in motion. This included the replacement of steam with diesel and electric trains. The electrification of lines is expensive to establish but is economical to run, so it has been provided on the very busy routes. Modernisation has extended to all branches of British Rail, highly sophisticated systems of signalling and mechanical loading and unloading of freight being particularly outstanding.

Today British Rail concentrates on three vital aspects:

(1) A fast passenger service from city centre to city centre which can compete with air services as regards speed, comfort and cost. This it hopes to do for distances up to 1000 km but more particularly in the range 200 to 500 km and at speeds of up to 240 kph with the Advanced Passenger Transport train.

(2) Local commuter traffic, which is bedevilled by the acute problem of uneven flow so that stations and trains are little used for much of the day but crammed to suffocation for short periods in the morning and evening rush hours.

(3) A freight service which is placing more emphasis on bulk movement between major centres. Increasingly this involves the movement of a block of specialist wagons carrying ore, oil, cars, chemicals or manufactured goods. Mixed freight headed for a variety of destinations results in frequent breaking up of the wagon line in shunting yards, an expensive and time consuming process. In the last few years the expansion of freightliner services with containerised commodities is providing a further extension of the freight service.

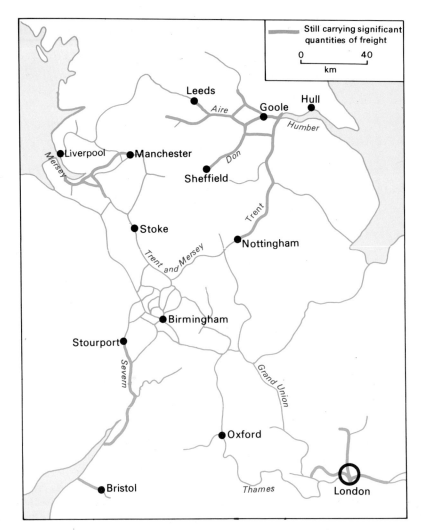

**Figure 6.7  Navigable waterways of the Midlands.**
The map shows the significance of canal systems linked to the four main estuaries. Much of the central system around Birmingham has been little used since the mid nineteenth century. In recent years there has been a rebirth of the use of canals especially for pleasure craft.

144

# INLAND WATERWAYS

Inland waterways had a great influence during the early stages of the Industrial Revolution. The major natural waterways of Britain present problems as routeways. Variability of flow, which influences the water level, excessive meandering and silting are three major handicaps. Consequently a canal network was developed particularly over the Midlands of England because natural waterways provided little scope for improvement and development. The construction of Britain's canal network provides some of the most fascinating aspects of our social and economic history. This pattern was established in a remarkably short period from about 1770 to 1800 and much of the construction was done using nothing more than spades and hand barrows. As the Industrial Revolution progressed the canal with its horse-drawn narrow boats could not cope with the demands of an ever increasing volume of traffic. From 1830 railways started to compete although it took some time before they made serious inroads into the proportion of traffic carried by canal.

# AIR TRANSPORT

For passengers travelling distances of over 500 km, air transport is certainly likely to be considered instead of rail. There is some movement of passengers by air over 250 km but the critical distance at which air travel becomes a viable proposition is very difficult to establish. Certainly as material standards of living rise there is a tendency for this distance to decrease. On the other hand the improvements in rail transport are likely to lead to intense competition. What blurs any comparison between air and rail transport for passengers is the time taken to reach the airport or railway station.

With regard to freight traffic, one thinks of perishable commodities such as flowers when out-of-season, or high value commodities such as bullion as principal forms of air freight. Although the vast majority of goods handled by British airports are international, the range of goods has increased considerably in recent years and engineering machinery, electrical goods and clothing are now the leading commodities by weight.

Heathrow is Britain's third port if value of freight is the criterion although in terms of volume it is relatively insignificant. Containerisation is a particularly valuable feature of freight traffic.

Ideally a new airport should be built on flat, well drained, cheap land where there is usually little fog, and near the large centre of population it serves. Obviously there are conflicts: the nearer to centres of population, the more difficult it is to find relatively undeveloped land. Airports also create problems for the local area, including noise and air pollution.

**Figure 6.8 Some Airports of the United Kingdom.** The size of the dots reflects the amount of both domestic and international traffic.

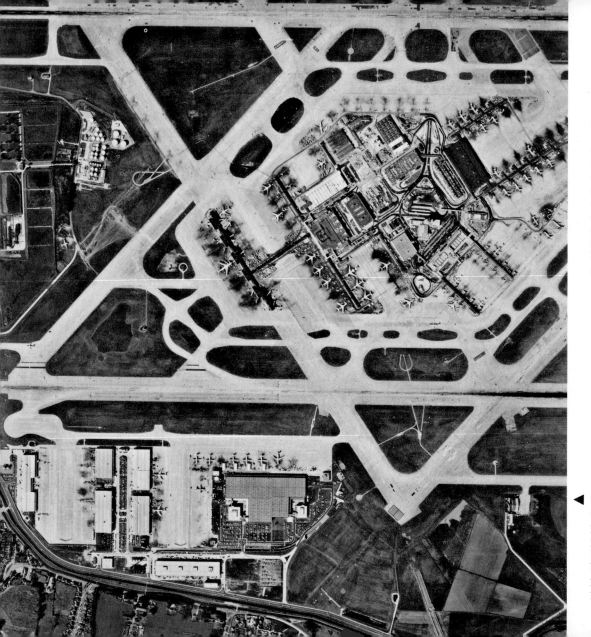

## EXERCISE

(1) How long does the rail journey from Glasgow to London take and what is the distance?

(2) Make similar calculations for the journey from Ardrossan to Guildford. How do the differences between the sets of figures illustrate some of the points made in the previous sections?

(3) Cost the journey by rail from Glasgow to London. How does this compare with the current air fare? How long does the trip by aircraft take and how much time do you think should be added on for the journeys from the airports to the city centres?

(4) What does the abbreviation APT stand for, and what effect will APT have on the competition between air and rail for the London–Glasgow run?

◀ What factors do you think should be considered when the site for a major airport is being selected? Illustrate your answer with references to the photograph on this page which shows a portion of London Airport, Heathrow. Consider availability of land and its use before development takes place, size of projected airport, relief and communication links with centres of population. (*Photo: Aerofilms*)

# Section 7
# RECREATION

# INTRODUCTION

During the course of the twentieth century there has been a marked improvement in the overall national standard of living. This improvement has progressed at an accelerated rate during the last twenty-five years. It is reflected in the rise in the purchasing power of wages: by that we mean that the rise in wages has been at a greater rate than the rise in prices. With more money the individual is in a position to provide himself with more material benefits and one of the first of these is likely to be a car. At the same time there has been a reduction in working hours and more and more families are able to enjoy long weekends away from their homes. Also there is a trend towards longer holidays and a growing number of people take more than one holiday during the year. Consequently there is an ever increasing amount of leisure time available and an enormous pressure on recreational and leisure facilities throughout Britain.

# HISTORICAL

Until the late nineteenth century holidays were for the very rich. There were seaside resorts such as Scarborough, Brighton and Bognor Regis but these attracted visitors rather for the benefit of their health than recreation. Towards the end of the nineteenth century an increasing number of

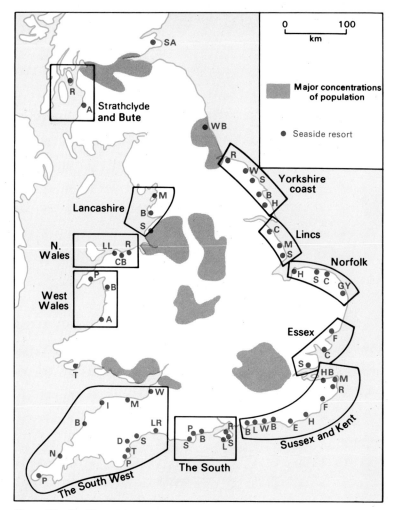

**Figure 7.1  Seaside resorts.**

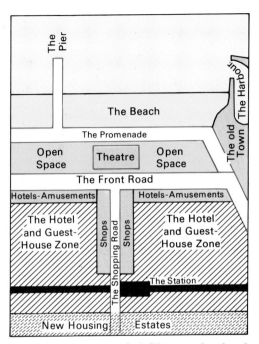

**Figure 7.2** **The anatomy of a holiday resort based on the functional structure of Paignton.**

people from the industrial regions of Britain began to take a seaside holiday. Few at that time had holidays with pay so they tended to go to the nearest resort available. In this way for example the centres along the north coast of Wales such as Llandudno, Colwyn Bay and Rhyl catered for holiday makers from industrial Lancashire.

**EXERCISE**
Where would holiday makers probably come from at the turn of the last century if they were at (a) Whitley Bay, (b) Scarborough, (c) Rothesay, (d) Skegness, (e) Southend?

As communications improved and the buying power of wages increased a greater number of holidaymakers went farther afield. During the first half of the twentieth century the more isolated south-west of England became popular and latterly in the third quarter of this century the 'package-deal' foreign holiday has proved attractive to many millions.

# THE SEASIDE RESORT

Accessibility had been one factor in establishing the early holiday resorts. Their popularity has been maintained or developed on the one hand by the exploitation of the physical advantages of the area and on the other by expanding the range of entertainment facilities. In the former case a sandy, gently sloping beach, a coastline of small bays, cliffs, rocky promontories and an attractive hinterland for inland visits are all factors. For many people, however, it is not the natural attractions but the man-made amenities such as the funfairs, children's play parks, coloured lights and stage shows which make a particular seaside resort attractive.

The major problem facing seaside resorts

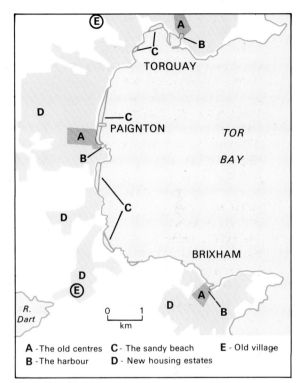

A - The old centres    C - The sandy beach    E - Old village
B - The harbour    D - New housing estates

**Figure 7.3** **Torbay.**

is the limited season, which creates a remarkably unbalanced demand for labour. This is partly remedied by using students on summer holiday and recruiting all available local labour, but even so this labour force must be added to by bringing in workers from elsewhere. On the other hand the long off-season period is one when the sea front presents a deserted, ghost-town appearance. Attempts to use the facilities, in particular the hotels,

149

for other purposes after the holiday period has met with some success in larger centres such as Blackpool, Brighton and Scarborough. These centres provide attractive rendezvous for all manner of conferences. Also as the length of the holiday season has increased, so the climatically more favoured resorts such as Torbay and Bournemouth can with reason boast that their season lasts from Easter to the end of October.

# THE RANGE OF RECREATIONAL ACTIVITIES

Catering for holiday makers at seaside resorts is only one aspect of the recreation industry. Although increasing numbers of foreigners, especially from north European countries, now visit our holiday resorts, particularly along the south and south-west coasts, the majority of overseas visitors are more interested in Britain's historic centres. London is inevitably the leading attraction both as a show-piece and a base for daily excursions, but towns such as Stratford-upon-Avon, Edinburgh and York attract more foreign visitors in the course of a year than any seaside resort. However, of more significance is the enormous increase in pressure on the use of all the recreational facilities available to the general public. These facilities have an enormous range in size, character, location and utilisation.

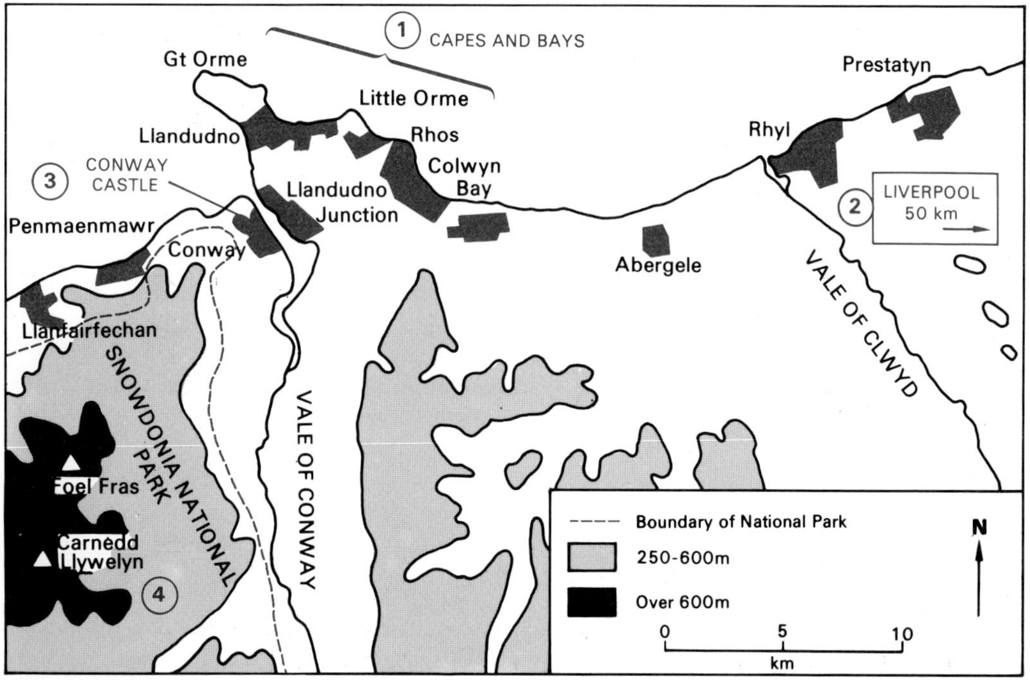

## EXERCISE

Consider the following list which includes various types of recreational facility, then answer the questions below.

   i) Historical site open to the public
  ii) A stately home open to the public
 iii) Forestry Commission land
  iv) Nature Trail
   v) Local recreational ground
  vi) Amusement park and funfair
 vii) Local public park
viii) Zoo and pleasure garden
  ix) Museum and art gallery
   x) Camping site
  xi) Golf course
 xii) Soccer, rugby, cricket and hockey pitches
xiii) Athletics track
 xiv) Tennis, badminton and squash courts
  xv) Swimming pool—heated and covered
xvi) National park

(1) Locate the nearest of each of these to your home. (There may be overlaps: for example many nature trails are on Forestry Commission land and many urban centres have combined sports

**Figure 7.4  North Wales.**
How do the references 1, 2, 3 and 4 on the map show evidence of the factors which have accounted for the growth of the holiday centres along the North Wales coast?

centres.) State how far away each of the locations is from your home.

(2) How much use is made of each of the locations? Is it ever overcrowded and when would this be?

(3) What other buildings and open spaces of a recreational nature can you think of which are near your home and have been omitted from the list?

(4) Examine critically the recreational facilities in your local area.

# THE NATIONAL PARKS

In 1950 the National Parks and Access to the Countryside Act came into operation and since then ten National Parks have been recognised. Within the boundary of the National Parks most of the land is privately owned and access is limited. The parks have towns, villages, farmland and forests within their boundaries but what is important is that these areas have great natural beauty and have been recognised as National Parks in an attempt to preserve this. Consequently there are restrictions on all forms of development, although you may think it unfortunate that there are eyesores in some of the parks. For example, Fylingdale Moor on the North Yorkshire Moors is dominated by the giant dishes of a radar tracking station while Milford Haven is the scene of massive oil refinery complexes.

In spite of the fact that the National Parks are in the more remote parts of England and Wales, some of their more accessible por-

tions are suffering from the effects of over-saturation by visitors. This can be seen at peak times in the form of crowded car parks and access roads jammed with vehicles, but more lasting is the ecological damage which is taking place; that is the damage to the natural environment which must result when countless numbers of tourists walk over a limited area of countryside.

The National Parks' authorities can make provision for visitors by providing car parks, shops, cafes, toilets, picnic areas and camping sites. The dilemma facing administrators is how to preserve the natural environment and improve facilities. Improvement both in amenities and access would surely encourage more people. The opening of the M6 motorway, for example, has made the Lake District easily accessible to many millions of people from Lancashire, Cheshire and the Midlands who want a day's excursion. Alternatively there need be no improvement so that additional visitors are discouraged. The answer surely is to improve facilities in more and more areas. In particular there is a need for more major park areas close to the densely populated parts of the country particularly in the Midlands and South-East. It is true that the government has control over other areas which are designated as being of outstanding natural beauty and there is a long-term intention to increase and enlarge on these.

In Scotland the problems associated with oversaturation of open spaces is not so great

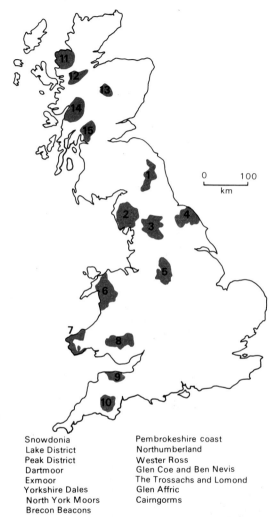

Snowdonia      Pembrokeshire coast
Lake District      Northumberland
Peak District      Wester Ross
Dartmoor      Glen Coe and Ben Nevis
Exmoor      The Trossachs and Lomond
Yorkshire Dales      Glen Affric
North York Moors      Cairngorms
Brecon Beacons

**Figure 7.5 National Parks and National Park Direction Areas (Scotland).**

1. Name each National Park from 1 to 15.
2. Choose three other areas of outstanding beauty which you think should be designated National Parks. Give reasons for your choice.

as in England because the densely populated areas are much more restricted. Nevertheless, the more accessible parts of the Highlands are beginning to experience some of the features of over-use which are all too familiar in the Lake District for example. The National Park Direction Areas of Scotland are shown in figure 7.5.

The National Trust and the National Trust for Scotland are two organisations which play a major part not only in acquiring historic buildings, ranging from castles to corn mills, but in opening up picnic sites and footpaths on the extensive tracts of land they own. The opening of the Pembroke coast footpath is an outstanding example of this. A number of National Trust properties, including Flatford Mill in Suffolk and Malham Tarn Field Centre, have been leased to the Field Studies Council.

## Local study: Beacon Hill, Charnwood Forest

This is a popular short-stay site (one to two hours) readily accessible from a number of East Midland towns including Nottingham, Derby and Leicester. There are hard-core and grass parking areas, toilet facilities and, at peak periods in the summer, a mobile refreshment bar. Weekend oversaturation is frequent during the summer months. The site is limited in size but there are attractive views over the Trent and Soar lowlands.

**Sample activity survey: Beacon Hill 13th September 1975, 1630 hrs**

Number of cars in the car park ................39
Activities of the people on the site:
    Sitting in cars.......................................... 7
    Sitting by the car, reading, taking a picnic ....................................................32
    Walking and sitting on Beacon Hill......58
    Playing ball games in the car park........ 9
    Flying kites ............................................. 5

## EXERCISE

Apply this survey to a similar site in your area.

The same pattern can be seen on other open spaces in the Midlands such as on Cannock Chase and Clumber Park in Sherwood Forest.

This topic of leisure and recreation is a complex one. Certainly we cannot all obtain the quiet and solitude we may seek, but we all now want and can afford recreational facilities which range over an enormous compass.

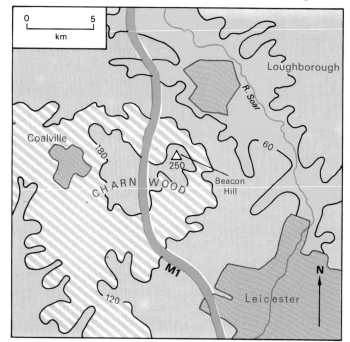

**Figure 7.6   Beacon Hill, Charnwood.**

River Thames, West London.

*(Photo: Aerofilms)*

Section 8
## POPULATION AND SETTLEMENT

# SETTLEMENT IN RURAL AREAS

In this section we are considering the pattern of settlement which emerges in those areas where the main economy is that of obtaining a livelihood from the land by farming or forestry. At the beginning of the Industrial Revolution over 90 per cent of the population lived in rural areas; today under 10 per cent do. Today the pattern of rural settlement is often blurred, sometimes by the introduction of a factory into the area and nearly always by the fact that many people who work in towns prefer to live in a rural community. It is the commuter element which so often has forced a village to lose its rural identity in the course of a few years.

## The pattern of settlement in a portion of the Midlands

This is a region of undulating relief with a surface of boulder clay, and the farming is mixed. Some farmers concentrate on livestock, specialising in sheep, dairy or beef cattle supported by fodder crops, others are only producing cash crops such as barley, wheat and some sugar beet. Others again have, within the single holding, a mixed farming economy of livestock and cash crop production. However, from the point of view of this study there is a uniform level of fertility over this region. In the region shown in figure 8.1, the land is not subject to flooding but in the past water was readily obtainable from shallow wells. It is difficult to distinguish physical factors in the region which might attract or repel people from the selection of a particular site.

Anglo-Saxon and Scandinavian settlements which grew up during the 'Dark Ages' were sited with due regard to the availability of necessities to make each community self-sufficient. Thus cultivable land for crops, pasture for stock to provide food and clothing, timber for fuel and building materials and a reliable supply of water were all necessary to serve the needs of the village. The limits of land serving a particular village community depended on a subtle balance. The needs of the village had to be satisfied so that there was a link between the size of the village and the potential productivity of the surrounding land. At the same time there was an economic limit. Farming of arable land 10 kilometres away from the village would mean perhaps three or four hours of walking a day. In practice most activities took place within 2 kilometres of the village.

In this region of the Midlands the settlements established during the seventh and eighth centuries were scattered with areas of woodland in between. In subsequent centuries there was an infilling to produce a landscape of tightly nucleated villages forming a regular pattern across the countryside. Later dispersed dwellings were added to the originally nucleated pattern.

## The pattern of settlement in a portion of the Fenlands

If you study the road pattern in figure 8.3 you can distinguish two markedly different patterns; one of straight roads, the other of irregular roads. This contrast reflects the different periods of settlement for the two regions. The area covered by the irregular pattern stretches in a series of sinuous and broken belts across the Fens. This land is slightly higher than that on which the regular pattern has developed.

During the Middle Ages much of the Fenland was marsh or waterlogged for portions of the year. Settlement sought firm land where foundations could support the building and provide dry homes. You may know of cases even today in which building land has been inadequately drained and newly constructed houses are suffering badly from floor damp and mildew on walls.

In the newly developed areas the regular geometrical patterns have been dictated by the drainage system of the major drains and feeder ditches. Here linear development along a road rather than a nucleation is general. Sometimes the development is a tight linear pattern, forming a village community which has a social unity. On the other hand where there is more general arable farming which supports a less densely populated area than the market garden concentrations, the settlement is more scattered.

The Fenlands, when drained, provide some

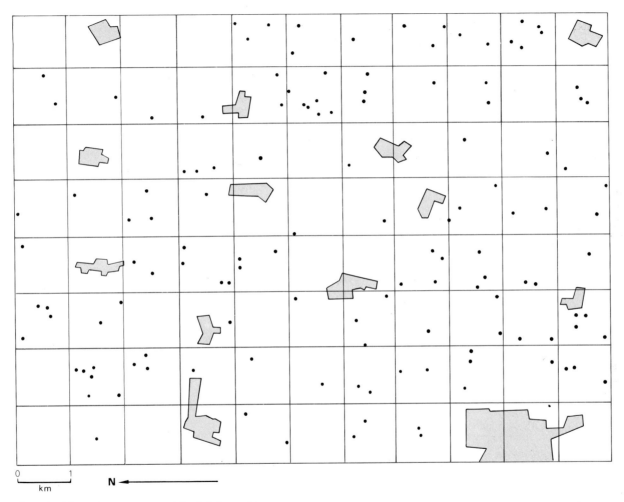

Figure 8.1 The settlement pattern of South Leicestershire.

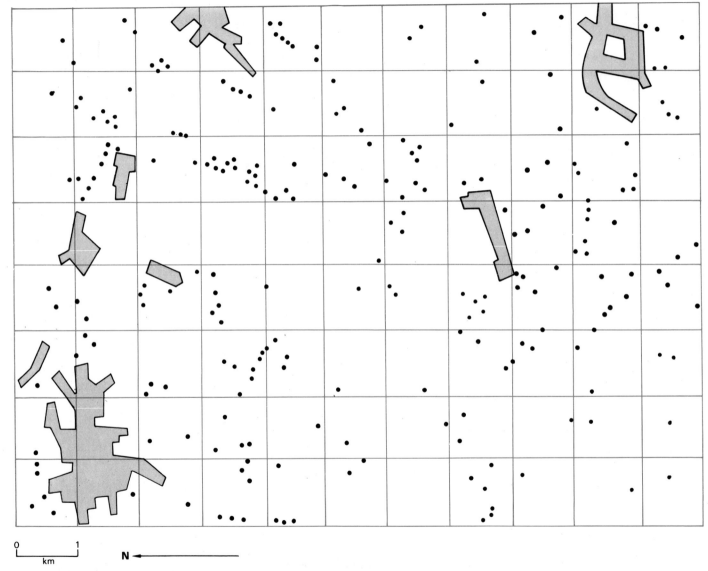

**Figure 8.2** The settlement pattern of a portion of the Fens.

of the richest agricultural land in the country. The economy is one of intensive arable farming including market gardening (refer to page 70). The size of holding is small and the density of rural population is high. In the long-settled areas, as farming intensified during the course of the twentieth century, more land was used for market garden crops. As a result there was an infilling of settlement and ribbon development along roads frequently linked one nucleation with the next.

## The pattern of settlement in a Yorkshire dale

Where farming potential is limited, fewer people can be supported by the land. At the same time, in upland areas where there are marked changes in relief, the sites available for settlements are often restricted. If you study the diagram of a portion of Swaledale (figure 8.4) you should notice that there are both positive influences which encourage the siting of settlements and negative influences which tend to repel settlement. At the same time notice the size and number of nucleated settlements in this area and make a comparison with the pattern in the other examples.

*Negative factors*  1. Steep slopes.
2. Exposed moorland covered by rough pasture.

3. The alluvial areas in the middle of the valley, seasonally waterlogged.

*Positive factors*  1. Better drained land at side of the valley providing firm foundations.
2. Nucleation where a tributary valley provides easier access to the upland (the largest nucleation occurs where there is a major tributary valley).

## EXERCISES

(1) List as many factors as possible mentioned in this chapter which can be locating factors in the siting of a village. You should be able to find at least ten such factors.
(2) Describe the pattern of settlement in figure 8.5 and use the information given in figure 8.6 to help account for this.
(3) For each of the four rural settlement maps (figures 8.1, 8.2, 8.4 and 8.5) indicate the proportion of squares which have evidence of settlement in them. What shortcomings has this method in comparing the density of settlement on the four maps?

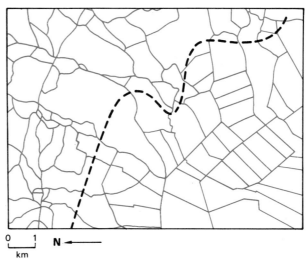

0   1
km   N ←

**Figure 8.3   Fens road pattern.**

157

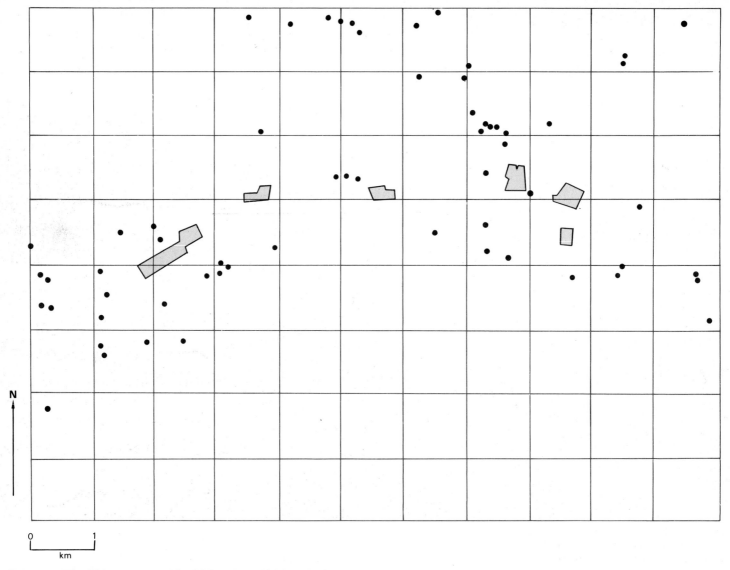

**Figure 8.4  The settlement pattern of Swaledale and the adjoining uplands.**

**A Yorkshire Dale.** Point out the features on this photograph of Wharfedale which are similar to those you
associate with Swaledale. (*Photo: Aerofilms*)

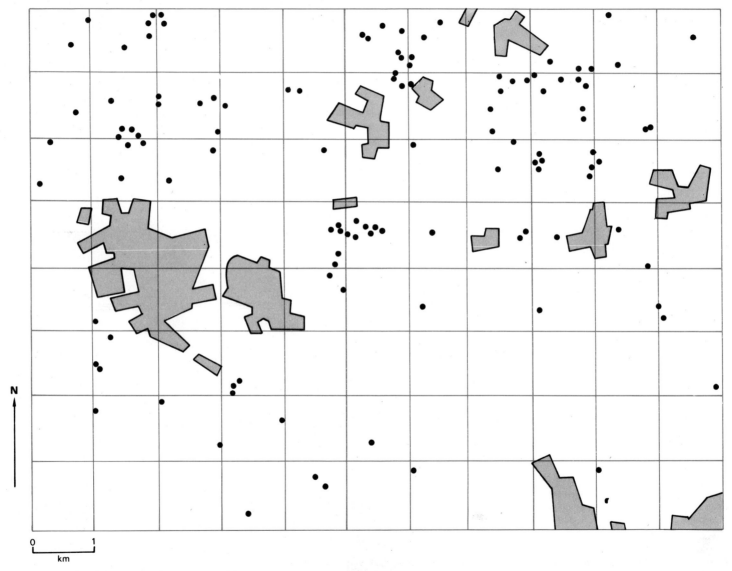

**Figure 8.5** The settlement pattern of a portion of Sussex.

## Depopulation in rural areas

We have seen how the density of the rural settlement pattern reflects the intensity of farming activity. It is likely that in areas of very low agricultural production the response will be a pattern of widely scattered hamlets and by contrast in regions of intensive cultivation there will be a high density of population with numerous villages and scattered farmsteads.

In the twentieth century many people have been attracted away from the regions of difficulty in Britain to the more prosperous parts. This has been the case in much of the highlands and islands of Scotland where isolation and a difficult terrain has discouraged particularly the young from remaining. Under these circumstances in such areas of depopulation the margins of cultivation retract to the edges of the coastal plain and wider, richer valleys leaving many of the settlements in the more remote, upland parts in ruin. (Refer to p. 62 on crofting settlements.)

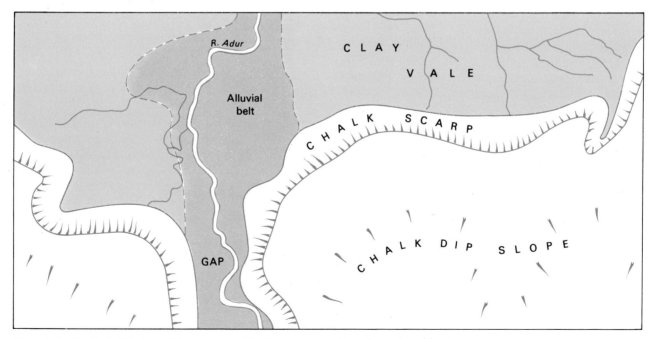

**Figure 8.6   The physical features of the area covered by the settlement pattern of a portion of Sussex.**

# URBAN SETTLEMENT

In the previous section we have been dealing with rural settlement. Within the pattern of villages there were some communities which took on a greater significance. This may have been because they were based around the castle of a feudal lord or because they became the venue for a periodic market. Such a settlement was likely to have some outstanding local geographical advantage which made it a focal point for the surrounding rural area. As it grew larger than its neighbouring villages it took on various functions other than being a grouping of homes for farm workers. One of the prime functions was to provide market facilities. Eventually such a favoured settlement became sufficiently important locally to rank as a town.

## The development of the Market Town

At the beginning of the section on rural settlement we noted that barely 10 per cent of the population lived in towns as recently as the mid-eighteenth century. Then most towns were, above all else, market centres. There was manufacturing but this was in small units and rarely dominated as it did later, in so many cases, with the coming of the factory system.

Today these long established market towns have frequently altered enormously in size and function but the degree to which these changes have taken place varies:

(1) The market town may have retained its original function without significant additions. That is, it remains a market town and expansion has been limited. Examples occur throughout the more rural parts of the country and include Salisbury, Hexham, Kendal, Carmarthen and Dumfries.

(2) The commercial function remains dominant but there are additional and significant functions, particularly manufacturing. Taunton, Norwich, Lancaster, Maidstone and Perth fit into this category.

(3) Many market towns have developed enormously since the Industrial Revolution as major manufacturing and service centres. Leicester, Nottingham, Northampton and Preston are examples.

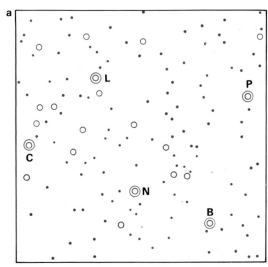

**Figure 8.7  Midland settlement patterns.**
(a) The actual pattern of settlement.
(b) The ideal pattern which should tend to develop when physical and economic influences are uniform over the whole area. The ideal situation suggests a 'hierarchy' of settlements, with large market towns, small towns of local importance and villages all at fixed distances forming a regular pattern.

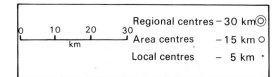

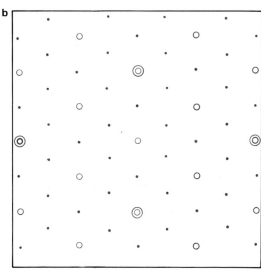

## The Industrial Town dating back to the early stages of the Industrial Revolution

Much urban development took place in the eighteenth and nineteenth century as many people from rural areas were attracted to the new factories, which they felt provided a greater degree of security, if nothing else. At the same time there began a period of rapid increase in the total population of the country. Villages on the coalfields mushroomed out into towns which sometimes merged into one another to create conurbations. In this way the West Midland or 'Black Country' conurbation was created.

## Coastal Towns

The great variety of coastal centres deserves a separate classification. As with inland settlement, coastal villages developed in favourable geographical locations affording shelter, both for the settlement and often for a harbour as well. Some more isolated coastal communities have retained their character; most have developed markedly different characteristics from former times.

(1) The little-developed coastal community maintaining its original function. For example along more isolated parts of the west coast of Scotland farming/fishing townships remain.

(2) The small fishing port which now caters

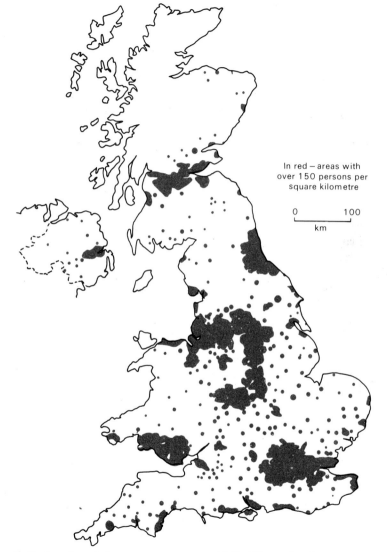

In red — areas with over 150 persons per square kilometre

0        100
km

**Figure 8.8   The distribution of population.**
Compare this map with that showing the distribution of coalfields. What main urban areas are *not* on or close to a coalfield?

163

for holiday makers as a dominant function. Many centres in Devon and Cornwall fall into this category.

(3) The seaside resort which developed as such at the end of the nineteenth century (refer to page 149).

(4) The naval base. Chatham, Portsmouth and Plymouth are examples which during the twentieth century have gradually extended their function, partly as a result of the decline in the nation's naval commitment.

(5) Ferry ports are those which provide limited facilities for short sea crossings. They include towns such as Harwich, Dover, Newhaven, Fishguard, Heysham and Stranraer. They have widened their function in recent years. To the passenger traffic, which shows a marked seasonal concentration, there is the expansion of freight transport, including container services necessitating additional facilities.

(6) Commercial ports vary in size and range of functions. During recent years the trend has been for various parts of the dock area to develop a specialised function, as will be seen in the special studies.

## New Towns

These are self-contained communities which have been established mostly since the 1946 New Towns Act. Normally such new towns have been based upon long-established villages to provide facilities during the early stages of expansion. New towns have been set up for one or more of a combination of reasons.

(a) To relieve overcrowding in nearby congested urban areas.

(b) To promote new industry in an area.

(c) To provide housing and amenities which are sufficiently attractive to the population to enable industry to expand and diversify rather than risk stagnation.

## The structure of towns

Although towns have evolved for different physical and economic reasons, certain functional areas are common to each one.

### The commercial sector

The concentration of routes at the centre of the town allows the maximum accessibility for the main commercial sector which develops there. The high rateable value of the land in the centre of a town is reflected in the type of shops, offices and services in this core area where banks, department stores and clothing shops develop. This is sometimes known as the central business district (CBD). The number and variety of concerns in the CBD will depend upon the size of the population served by the town and this will partly be influenced by the proximity of towns of comparable or greater size. For example Marks & Spencer usually establish stores in towns whose sphere of influence includes 60 000 people, although

towns which attract a large tourist population may be exceptions.

There are four methods which can be used to delimit the boundaries of the CBD. Use two or three of the methods to obtain a more precise result.

(1) A ground-floor land-use survey plotting the information on a large scale map such as the 25 or 50 inches to 1 mile plan.

(2) A first- and second-floor survey giving the use of the building at that level.

(3) A pedestrian-flow count based on a scatter of survey points so that a flow chart for pedestrian movement can be built up. Maximum movement will be in the CBD.

(4) A comparison of rateable values of premises taking into consideration the total floor area or the frontage of the building. A record of rateable values is available for public inspection at the relevant council offices. However, extracting the information can be a lengthy process and is only useful when compared with some measurement of the area concerned.

Figure 8.9 illustrates a survey of the ground-floor use in the CBD of Taunton. Retailing outlets such as car showrooms and some furniture shops, which require large areas of floor space, are in many towns outside the CBD; the same is true of businesses with a relatively low turnover in comparison with the national chain stores. In Taunton most food shops lie outside the CBD and tend to have a more local clientele. A study of first- and second-floor use for commercial

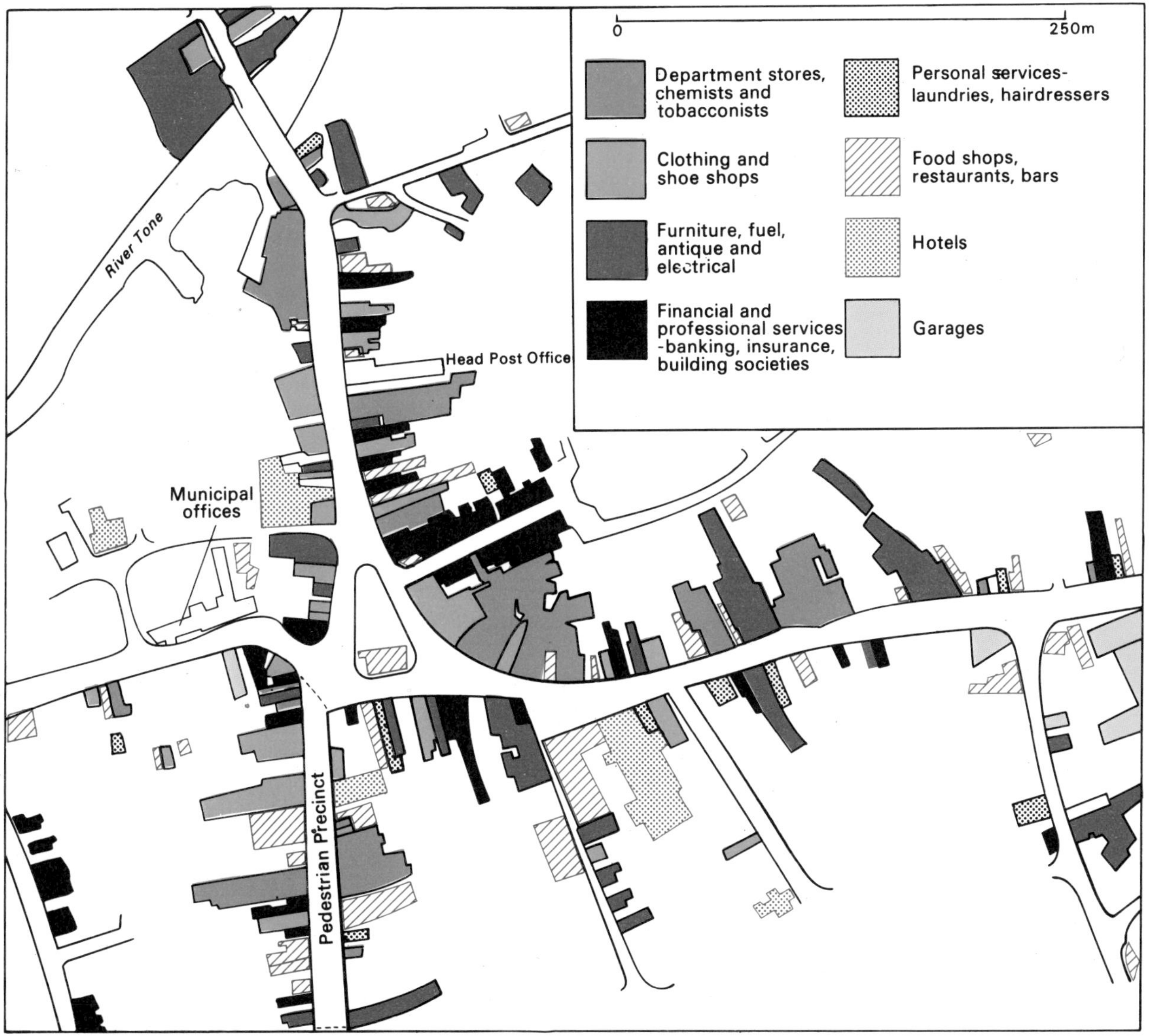

0
250m
Department stores, chemists and tobacconists
Personal services-laundries, hairdressers
Clothing and shoe shops
Food shops, restaurants, bars
Furniture, fuel, antique and electrical
Hotels
Financial and professional services -banking, insurance, building societies
Garages
River Tone
Head Post Office
Municipal offices
Pedestrian Precinct


**Figure 8.9 Central Taunton—groundfloor land use.**

purposes shows that such use occurs mainly within the CBD.

The boundaries of this commercial area can change. In Taunton the partial redevelopment of High Street with its associated extension of nearby parking facilities has influenced the movement of pedestrians in the town and subsequently brought a new area into the CBD.

**The industrial sector**
There may be more than one industrial sector in a town. Such a situation reflects the differing locational factors affecting industry over a period of time. Towns which grew as a result of the first stages of the Industrial Revolution usually have remnants of their early industrial growth close to the commercial centre. As industry expanded in the nineteenth century, it did so along the main lines of communication, particularly the railways. Much of this well established industry is interspersed with older housing. In the twentieth century changes in transport methods and the greater reliance on road transport have made these industrial areas less suitably located than they once were. There has been a tendency for new manufacturing plant to concentrate in specialised industrial sectors away from the inner urban area even if there is abundant land there ready for redevelopment: the crucial factor has been the value of uncongested access to plant for raw materials, finished goods and employees as well.

**The residential sectors**
These may be classified according to age or cost of housing. The innermost residential areas are the oldest. Since in most industrial towns they were built for factory workers, proximity to the place of employment was important when passenger transport was limited. In some towns many of the oldest houses have been pulled down and replaced by modern dwellings; often these are flats which can accommodate the same number of

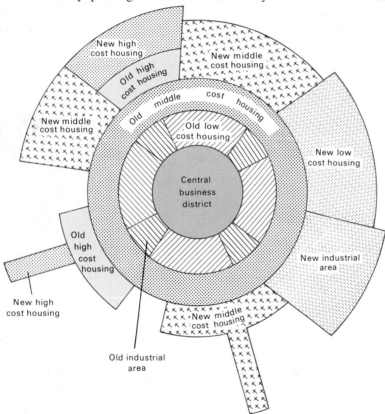

**Figure 8.10 The growth of a town.** The diagram shows the sector development which might occur as an industrial town grows.

166

people as the old terraces and open spaces are created between the buildings.

In larger towns where redevelopment has not taken place the central residential areas are often referred to as 'twilight zones'. They are characterised by low cost, high density dwellings. As people become more mobile, the residential areas have extended outwards, the density of housing tending to be reduced the farther out one gets. This has led to urban sprawl typical of so many towns today. London is the most outstanding example and details of its expansion will be dealt with below.

On the edge of a town lie the villages and more isolated housing developments from which people nowadays are prepared to commute to work. Unless the expansion of the town outwards is checked these villages may well be the next to be engulfed by the urban sprawl as others have been in the past.

# CARMARTHEN

Carmarthen, situated on the west bank of the river Towy, has always been an important centre in south-west Wales. Its earliest origins appear to date back to Celtic times when a hill fort was constructed on a rise adjacent to the river. Later both the Romans

Using information on the map and the photograph, ► describe and account for the site and form of Carmarthen. (*Photo: Aerofilms*)

and Normans recognised the strategic value of Carmarthen, which functioned as a port besides controlling the river crossing and routes westward. Although the river is now silted up and only serves pleasure craft and fishermen, Carmarthen's position as a route centre is reflected in its status as the administrative centre of Dyfed and the principal market town for the surrounding area.

Pastoral farming, with particular emphasis on dairying, thrives on the rich alluvial soils of the Towy Valley. With its extensive agricultural hinterland, Carmarthen supports a large cattle mart, dealing with over 120 000 head of cattle a year. At present this market is situated in the centre of town, but shortage of land for expansion will soon force it to transfer to the outskirts.

Carmarthen has not witnessed the phenomenal growth characteristic of some neighbouring towns in South Wales. As it is situated away from the coalfield, many people left Carmarthen in search of employment in the mines and developing industries during the late nineteenth century. The town's peripheral position has continued to discourage any large industrial development. Firms have established bases there, but often only to help distribute their products. A large creamery employs a considerable labour force, but most of the working population are engaged in the service sector. Recognising the problem of attracting industry, the district council has earmarked sites within the town's boundary for industrial estates. With

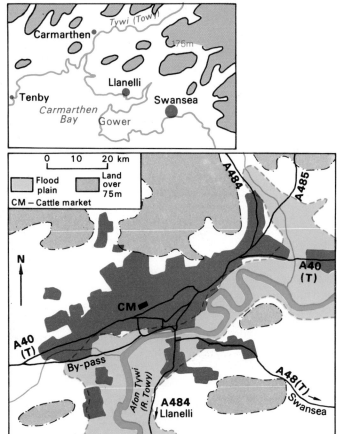

Figure 8.11 Carmarthen.

the advantages of development area status it is hoped to encourage light industry to this part of Wales.

Besides acting as an administrative and market centre, Carmarthen also provides shopping facilities for the surrounding area, and serves as a recreational and cultural centre.

# SHEFFIELD

Sheffield is an example of a town that has grown as a direct result of industrial expansion. Before coal became important as a source of energy Sheffield was already a significant manufacturing centre. Local supplies of coal measure ironstone, charcoal from nearby woodlands, water power available from the streams that broke through the coal measure scarp and grindstone material from the millstone grit area to the west had promoted this early industry. Sheffield has been noted for its cutlery since the Middle Ages and its reputation was enhanced when Flemish immigrants brought their skills to the area in the sixteenth century. Today this particular industry continues to produce a variety of products ranging from surgical instruments to razor blades and scissors. The production of silver ware is an off-shoot of the cutlery trade.

However, from the late eighteenth century the growth of the town has been associated with the expansion of other branches of the steel industry, particularly the manufacture of high-grade steel alloys and special steels. The Don Valley, downstream from Sheffield, provided the sites for this industrial development as well as a line of communication. The river was dredged and canals built to link Sheffield with Hull and the sea, as already imported high-grade iron ore was needed in place of the inferior quality local ore. During the nineteenth century the town spread out

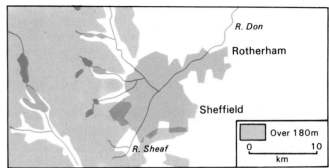

from its nucleus at the confluence of the Don and Sheaf up the sides of the valleys and on to the edge of the moorland to the west.

Sheffield has continued to expand its production of special steels. Characterised by their resistance to high temperatures and their strength, these are used in the production of machine tools and castings for turbine engines. These and other alloy steels such as stainless steel reflect the emphasis on the production of small quantities of a high-value finished product, a response to Sheffield's inland position. In fact no pig iron is produced here, considerable reliance being placed on scrap metal.

As in the case of other towns that developed early in the industrial era, Sheffield has been faced with the problem of redeveloping its central residential areas. Rather than rehouse people elsewhere, large blocks of flats have been built in place of the old tenements. Sheffield, with a population

Figure 8.12   Sheffield.

of 500 000, has also emerged as a shopping and cultural centre for the north Midlands and south Yorkshire area. Much of the commercial district, which was badly damaged during the war, has been rebuilt.

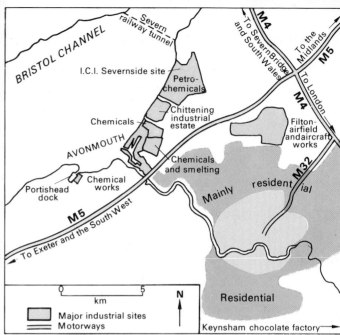

**Figure 8.13 Bristol and Avonmouth.**

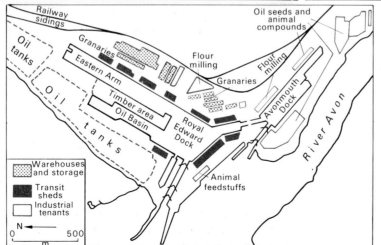

**Figure 8.14 Avonmouth Docks.**

**1.** Why have the Avonmouth Docks, once the outport for Bristol, overshadowed the old docks in the centre of the city?

**2.** Describe the functions of Avonmouth Docks today.

**3.** Why should certain types of industry develop along the Bristol Channel upstream from Avonmouth?

# THE PORT OF BRISTOL

The original port developed in the early Middle Ages at the junction of the rivers Frome and Avon, where a peninsula slightly above the surrounding land provided the site, 13 km up the Avon. In Norman times it expanded as the nation's trade also developed. In the sixteenth century Bristol's trade with North America, the Caribbean and West Africa began, as did the city's staple industries, of which the most outstanding relied on imported cocoa and tobacco. By the eighteenth century, with the increasing size of ships the restrictions of the river site were felt, and movement mostly occurred at high tide. Nevertheless, in the early nineteenth century the 3 km long 'New Cut' was dug out of solid rock to divert the Avon, and the 'Floating Harbour', which extends for 5 km along the Avon, was created. During the latter half of the nineteenth century, with developing interests in our colonies and the consequent increase in our volume of trade, there were great incentives for further expansion. During this time it was Avonmouth which started to widen the range of imported goods; it became a major importer of grain and soon afterwards of frozen meat, and in both cases this necessitated the introduction of specialist storage facilities. In 1908 the construction of the Royal Edward Dock at Avonmouth reflects the abandonment of further schemes on the Avon itself. Future developments became

## Industries associated with the port

| Imports | Origin | Industrial concern |
|---|---|---|
| Chemicals and fertilisers | Liquid phosphorus (Newfoundland) | Albright and Wilton (Portishead) |
| | Phosphate of lime (Nauru) | ICI and Fisons (Avonmouth) |
| | Potash (USSR, Spain, Israel, Holland) | ICI (Hallen) and Fisons (Avonmouth) |
| | Nitrates (Holland) | ICI (Hallen) and Fisons (Avonmouth) |
| Feeding stuffs | Rice, bran (India, Pakistan, Guyana) | Various processing mills at Avonmouth Docks |
| | Fishmeal (Scandinavia, Iceland) | |
| | Oilnut meal (Denmark, India) | |
| | Oilseed meal (Italy, India, Pakistan, Algeria) | |
| Tobacco | USA, India and Africa | W. D. & H. O. Wills Ltd., Imperial Tobacco Company (Bristol) |
| Cocoa | West Africa | J. S. Fry & Sons (Keynsham) |
| Lead and zinc | Lead (Australia, Canada, Peru) | Imperial Smelting Corporation (Avonmouth) |
| | Zinc (Australia, Canada, S. W. Africa, Yugoslavia) | |
| Timber and woodpulp | Hardwoods (W. Africa) | Various timber companies around the city docks |
| | Softwoods and woodpulp (Scandinavia, USSR, Poland, Canada, Brazil) | |
| Grains | Wheat (Europe, Canada, USA, Australia, Argentina) | Mills at Avonmouth |
| | Maize (USA) | |
| Petroleum | From refineries in South Wales, Fawley and the Thames Estuary | For general distribution and the CEGB Power Station (Portishead) |

concentrated on either side of the mouth of the river. In 1921 the Oil Basin was finished at Avonmouth and in 1928 the extension of docks to include grain berths also involved the introduction of suction equipment to improve the handling of grain. Meanwhile industrial development associated with the dock was rapidly expanding within the port

171

boundary as well as nearby. The table gives some idea of the range of this activity, although it deals only with the major concerns.

In recent years further developments employing new handling techniques have continued and the Oil Basin has been rebuilt. The revolution in freight handling by containerisation is reflected in the provision of new berths to handle this traffic.

# CRAWLEY

Crawley is one of a series of new towns built in a ring of 40 or 50 kilometres radius from London. These towns were established with the aim of relieving overcrowding within the capital. Crawley is based upon a long-established village of that name, the purpose being to give the town some identity in the early stages of development. The town was planned and built by a development corporation which began work in 1947 and was wound up in 1962 when much of the scheme was complete.

As you can see in figure 8.16, the plan consists of distinct residential areas, an industrial zone and a central commercial sector. The eleven residential areas are separated from each other, often by open spaces and always by main roads. Each area has been designed to be self-contained to a degree; for example they each have a primary school, shopping facilities, a church and community

meeting place. Three-quarters of the housing within Crawley has been built by the development corporation. Industry is concentrated in a single sector.

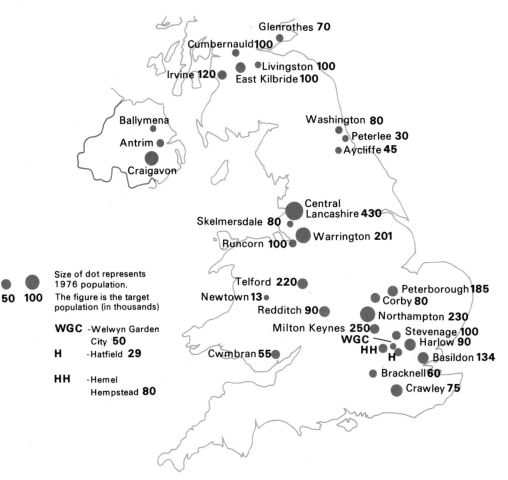

Figure 8.15 New Towns.

## EXERCISE

What do you think the planners needed to consider when locating the industrial sector?

The industrial zone has over one hundred separate firms concerned with a variety of light industry. Although the engineering group is dominant, there are also food processing, wood working, pharmaceutical and printing concerns.

The central commercial sector incorporates the old High Street portion of Crawley village with a new shopping precinct based around its square, and this now functions as a centre not only for Crawley itself but for the surrounding rural area.

Such a planned new development appears to create an ideal situation and yet there have been many problems. Londoners have been reluctant to move out to Crawley despite its better housing facilities. In fact many people who have settled in Crawley have come from other parts of the country. The housing areas have been criticised for their uniformity and cold, impersonal appearance and many people who work in the industrial sector travel in from outside the town. The age structure of Crawley's population is limited, with dominantly young married couples, and since they have mostly arrived in the area in recent years, there has been little time for them to build up any allegiance to the town.

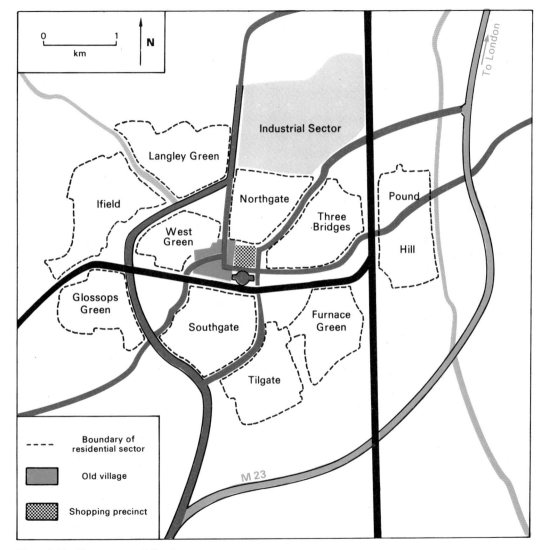

**Figure 8.16  The structure of Crawley.**

The photographs are of portions of Crawley (left) and Salisbury. **1.** How can you identify the old village on the Crawley scene? **2.** Compare the two towns using evidence on the photographs. (*Photos: Aerofilms*)

# ABERDEEN

Aberdeen, with a population of 210 000 is the largest settlement north of the river Forth. Owing to its focal position in the north of Scotland the city has a much bigger sphere of influence than is suggested by its function as the administrative centre for the North East region. The extension of its hinterland has resulted from its coastal position and port facilities. Sea links are maintained with the northern isles and more recently Aberdeen has become the mainland base for North Sea oil companies.

The site of Aberdeen is largely confined between the rivers Don and Dee. Originally two separate settlements existed; one, on the edge of a meander scar of the Don, was an ecclesiastical centre, the other, on the banks of the Dee estuary, concentrated on fishing and trading. A road linking the two settlements ran along the crest of a moraine to avoid the less well-drained ground. As the population increased in the nineteenth century so the intermediate area became settled.

The Dee estuary site was by no means ideal for a port. The presence of sand banks and a strong ebb tide hindered navigation. However, the Don offered little potential as the river passed through a rocky gorge before it entered the sea. Reclamation carried out at the beginning of the nineteenth century shifted the channel of the Dee southward and provided valuable land for industrial and commercial development and for building quays.

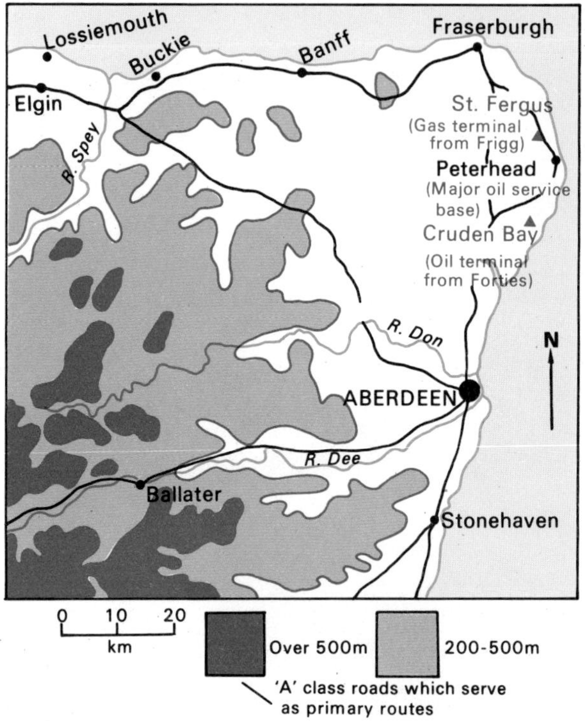

The port is primarily concerned with accommodating general cargo vessels, fishing and shipbuilding. Aberdeen is a leading centre for white fish in Britain. The major fishing grounds of the North Sea, the Faroes and Iceland are on the doorstep, although there has been a tendency to concentrate on the middle waters and leave the more distant areas to vessels from Hull and Grimsby. The fish wharf and fish market stand alongside the Albert Basin as do processing plants connected with the industry. The railway is nearby which facilitates the speedy transport of fresh fish to the south. Shipbuilding developed at Aberdeen partly because of the extensive trade carried out by the port. Nowadays very few ships are built, the yards concentrating instead upon the repairs and maintenance of trawlers and supply vessels for the oil rigs.

**Figure 8.17  The position of Aberdeen.**

Prior to the discovery of oil in the North Sea, industry in Aberdeen had not developed to the same extent as in towns in the Central Lowlands. Textiles, papermaking, marine engineering and food processing were the principal manufacturing industries. For a long time Aberdeen was renowned for the quarrying and polishing of granite, which is mainly used now to face buildings, and in the manufacture of paper-making rollers and large concrete pipes.

The oil industry has based itself on Aberdeen in particular because the town is already established as the major administrative, medical, communication, service and educational centre for the North of Scotland. This new industry has provided an enormous impetus for further growth. There has been considerable investment in new offices, warehouses and factories and an increased demand for industrial land. Existing industrial estates such as East Tullos have rapidly been filled and new sites have emerged to the north, particularly around Dyce airport, and to the south of the town.

Industries new to Aberdeen include the manufacture of a wide variety of articles needed on an oil rig and it is expected that some of these items will soon be exported to Europe and the Middle East. Many firms have established branches in Aberdeen to promote the supply of their goods, as in the case of Philips and Pye who manufacture electronics equipment. Established firms in the area have also been encouraged to

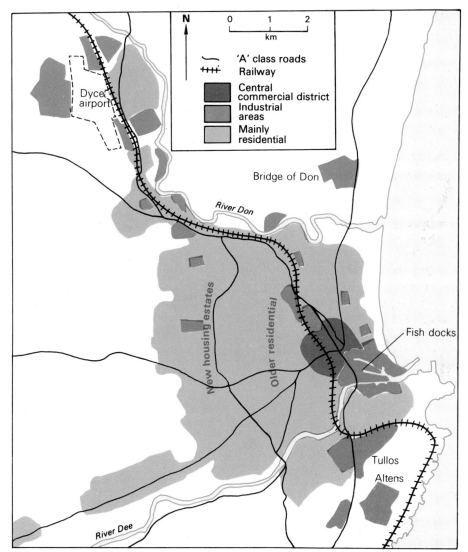

Figure 8.18 Aberdeen.

branch out into the manufacture of goods for the oil industry.

New developments have been initiated at the port. Besides the provision of roll on/roll off facilities and a new fish market and associated wharves, six integrated oil-field supply bases have been constructed. Storage facilities at the docks are vital and four of these bases are used exclusively by oil companies.

Many of the services needed by the oil industry are already provided. The University and Institute of Technology already offer specific courses in petroleum exploration and off-shore engineering. An added bonus has been Aberdeen's fishing industry which has made available many marine services and considerable expertise. In turn, the recent growth stimulated by North Sea oil is bringing about an expansion of services already provided. Dyce airport now has Britain's largest heliport. New hotels are being built and the shopping area is developing. This process will continue and is likely to lead to the provision of more commercial facilities and an even wider selection of cultural and recreational amenities.

# GREATER LONDON

London is not only the political capital of the country, but the first port and the leading centre for industry, finance, commerce, culture and entertainment. This concentration of major functions in one centre creates problems with no easy solutions.

## The growth of London

Roman Londinium was built on a site where two gravel mounds, rising above the surrounding marsh, provided a firm, dry foundation for buildings and also a suitable crossing point of the river. The two mounds, Ludgate Hill and Cornhill, are now the site of the City. Further west again, on a pocket of gravel, a royal residence and cathedral were erected in Saxon times at Westminster. The Normans, who constructed several strongholds around the settlement, including the Tower of London, provided the security for the town to expand.

As London has developed, different functions have tended to concentrate in separate sections of the town. The heart, the City of London, continued its business and commercial function. Below London Bridge, Thames-side wharves sprung up to cope with shipping; in this area, known as the East End, small factories and houses crowded together. To the west, beyond Westminster, fashionable residential areas such as Bayswater and Kensington grew up.

In the nineteenth century with improvements in transport culminating in the system of underground railways after 1890, people working in London were able to live farther afield and travel in each day. The relentless spread of housing engulfed neigh-

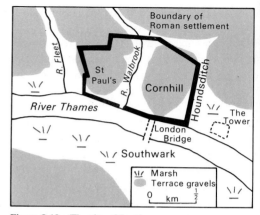

**Figure 8.19  The site of London.**

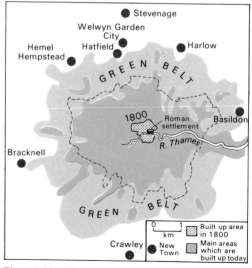

**Figure 8.20  The growth of London.**

bouring towns and villages such as Highgate, Finchley, Hendon, Harrow, Acton, Ealing, Twickenham, Kingston, Wimbledon, Croydon and Greenwich, these old centres being the nuclei for the new suburbs. At first the new suburbs developed on the better drained gravel terraces but with pressure on the land the low-lying clays were also built over. During the twentieth century this expansion has continued. As London snowballed creating a great reservoir of labour, and a large market in itself, and as industry no longer benefited from a position close to coalfields, so the London area became an ideal zone for industrial development. Manufacturing is concentrated in sectors around the city.

## The structure of London

Figure 8.22 illustrates in a generalised form the main functions of the various parts of London.

### The City

The City covers about 250 hectares but within this is concentrated the headquarters of the main commercial and financial groups in the country as well as the Bank of England, the Stock Exchange, Lloyds (for insurance) and Fleet Street, the centre of publishing and newspapers. Wholesale food markets including Smithfield for meat and Billingsgate for fish are also located here, but they are most unsuitable sites now. There is very little residential property.

### Westminster

This includes not only the Houses of Parliament but also government offices; many foreign embassies are nearby.

### The West End

The West End is the shopping, theatre and hotel centre. Here are also offices of many leading industrial enterprises. It is a high-class residential area.

### Port of London

The Port first developed in the Pool of London immediately below London Bridge. Here there is a tidal range which varies between 4 and 7 metres. With larger ships enclosed docks, where a certain depth of

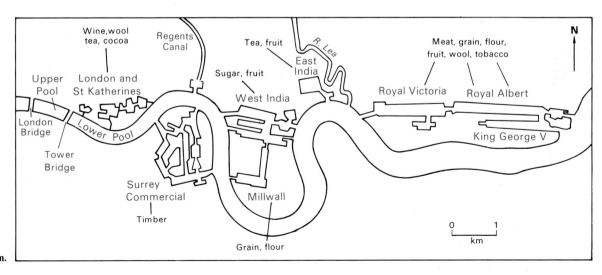

**Figure 8.21   The Port of London.**

179

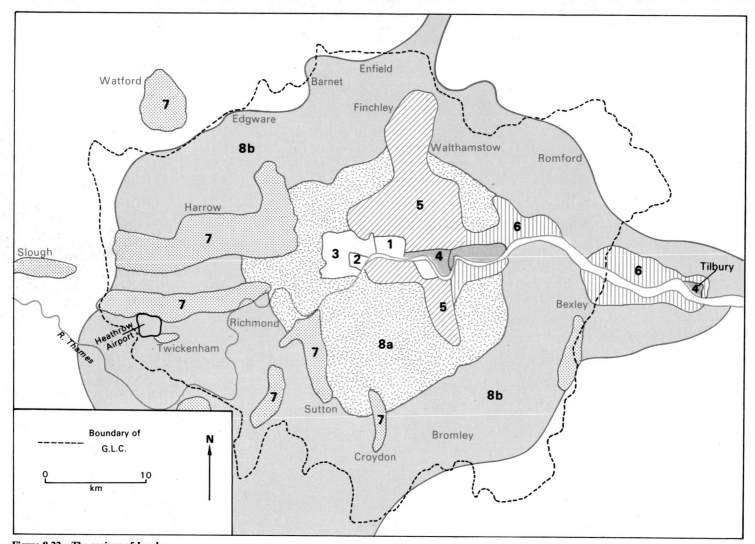

**Figure 8.22　The regions of London.**
Identify the areas labelled 1 to 7. 8a is the inner residential zone and 8b the outer dormitory zone.

**The Barnet area of North London. 1.** Draw a map of the area covered by the lower two-thirds of the photograph, marking in main roads, open spaces, industrial and residential areas. **2.** Comment on the pattern of land use for the whole area of the photograph. (*Photo: Aerofilms*)

181

water could be ensured, became necessary, and at the same time provided more wharfage. The alluvial ground bordering the river provided the site for the dock system, and such a soft type of rock presented no excavation problems. Since 1908 the docks have been controlled by the Port of London Authority.

### The inner industrial belt

This industrial zone first developed immediately to the north of the docks, but the types of industries were influenced more by the local market. Consumer goods such as clothing, boots and shoes, electrical goods, furniture and plastic household ware are the dominant products. The traditional small manufacturing units remain but newer, larger factories are becoming established especially where the belt is expanding northwards up into the Lea Valley.

### The lower Thames industrial belt

Industries which require bulky raw materials of low cost in relation to weight are at an advantage when ships or lighters can unload directly on to the factory site. Consequently there is a concentration of such industries along the banks of the lower Thames. Plants include sugar refineries, flour mills, leather tanneries, paper mills, factories producing chemicals, soap and detergents as well as power stations. The Ford Motor Company is also situated here. (See page 115).

### The industrial outskirts

As central London has been saturated with development new industry has sprung up on the outskirts, for example along the Grand Union Canal and Brent Valley, the A4 from Brentford and between Croydon and the Thames. These industries depend on road transport, hence their concentration on arterial routes. The character of such factories varies considerably, but they are generally larger than those found in the inner industrial belt. However, they also tend to produce consumer goods which are relatively valuable in proportion to their weight; for example clothing, food, furniture and electrical appliances.

### The dormitory areas

Elsewhere vast areas of housing exist, interspersed with parks and open spaces. Shop-lined streets mark the position of old villages and towns engulfed by suburbia. Scattered pockets of industry do exist.

## Trends

The chief concern for future planning is to restrict the size of population and the outward sprawl of building. Many people have been attracted to the London area from other parts of the country especially from those regions with a history of above average unemployment. This has been a feature of the last forty years because of greater job opportunities in London. To limit this ex-

pansion and restrict the growth of the population a 'Green Belt' was created around the city in which development is restricted, and centres were established outside this belt. Some of these centres are expansions of older towns such as Luton and Basingstoke; many are new towns, including Crawley (study figure 8.16). Other means of restricting growth have been to decentralise government departments, relocating them away from London, and to control the building of new offices and factories within London. This, it is hoped, will limit the increase in commuters who daily travel to London from the outer suburbs and areas up to 100 km away.

## EXERCISES

(1) Describe the site of London.
(2) Why did the city develop there?
(3) Consider the distribution of manufacturing industry in London.
   (*a*) Classify the types of industrial area.
   (*b*) What have been the influences on the establishment of industrial areas in the London region?
(4) What advantages has London which have made it the leading port of Britain?
(5) What is the purpose of:
   (*a*) a green belt around London.
   (*b*) New towns in a ring outside the green belt?

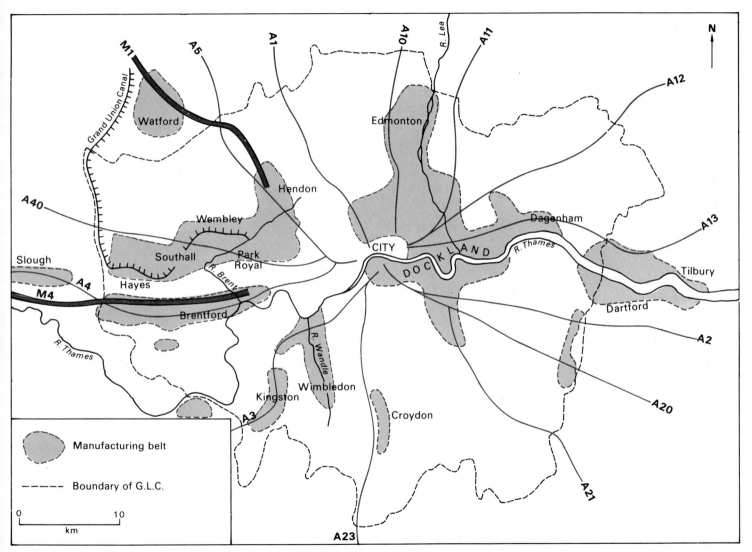

**Figure 8.23 Industrial zones and main routes—London.**

# INDEX